Studies in Systems, Decision and Control

Volume 150

Series editor

Janusz Kacprzyk, Polish Academy of Sciences, Warsaw, Poland
e-mail: kacprzyk@ibspan.waw.pl

The series "Studies in Systems, Decision and Control" (SSDC) covers both new developments and advances, as well as the state of the art, in the various areas of broadly perceived systems, decision making and control- quickly, up to date and with a high quality. The intent is to cover the theory, applications, and perspectives on the state of the art and future developments relevant to systems, decision making, control, complex processes and related areas, as embedded in the fields of engineering, computer science, physics, economics, social and life sciences, as well as the paradigms and methodologies behind them. The series contains monographs, textbooks, lecture notes and edited volumes in systems, decision making and control spanning the areas of Cyber-Physical Systems, Autonomous Systems, Sensor Networks, Control Systems, Energy Systems, Automotive Systems, Biological Systems, Vehicular Networking and Connected Vehicles, Aerospace Systems, Automation, Manufacturing, Smart Grids, Nonlinear Systems, Power Systems, Robotics, Social Systems, Economic Systems and other. Of particular value to both the contributors and the readership are the short publication timeframe and the world-wide distribution and exposure which enable both a wide and rapid dissemination of research output.

More information about this series at http://www.springer.com/series/13304

Christina N. Burt · Louis Caccetta
Editors

Equipment Selection for Mining: With Case Studies

Springer

Editors
Christina N. Burt
Department of Mathematics
and Statistics
The University of Melbourne
Parkville, VIC
Australia

Louis Caccetta
Department of Mathematics
and Statistics
Curtin University
Bentley, WA
Australia

ISSN 2198-4182 ISSN 2198-4190 (electronic)
Studies in Systems, Decision and Control
ISBN 978-3-030-09444-7 ISBN 978-3-319-76255-5 (eBook)
https://doi.org/10.1007/978-3-319-76255-5

Printed on acid-free paper

This Springer imprint is published by the registered company Springer International Publishing AG part of Springer Nature
The registered company address is: Gewerbestrasse 11, 6330 Cham, Switzerland

Preface

The mining industry contributes significantly to the health of the worlds' economy. Indeed, the total annual revenue generated by the industry worldwide has exceeded \$500 billion US for the past 6 years. Over this time, the net profit margins have decreased from 25% in 2010 to 4% in 2016. The significant global economic uncertainty together with declining trends in average ore grades, declining market prices, increasing mining costs, and the complex regulatory, environmental and safety restrictions in which the industry must operate is contributing to this decline in profitability. Consequently, the economic viability of the modern-day mine is very highly dependent on careful planning and management. This, of course, presents enormous opportunities for the application of cutting-edge optimisation technology. Optimisation techniques have been successfully applied to resolve a number of important problems that arise in the planning and management of large and complex mines. Applications that are well documented include ore-body modelling and ore reserve estimation, optimal pit design, optimal production schedules, optimal blends, effective equipment selection, utilisation and maintenance, efficient mine site rehabilitation and a range of transport and logistics issues. This book focusses on the important truck–loader selection problem.

The truck–loader selection problem is that of selecting a fleet of trucks and loaders for use in extracting ore and waste throughout the life of the mining operation. The cost of the truck and loader fleet has been estimated as being up to 55% of the total cost of the operation making the purchasing and maintaining the correct combination of trucks and loaders critical to the economic viability of a mining operation. For a new mining operation, setting up the initial fleet requires a large purchase of trucks and loaders for the removal of ore and waste as specified in the mine plan. The effective management of this fleet requires continual changes to this fleet as the mine plan progresses. Note that fleet equipment has a life cycle of around three to five years, whilst the life of the mine may well exceed 50 years. Prior to our work, the methods used for determining the trucks and loaders to use in a mining operation largely relied on the experience of specialist consultants with computational methods usually restricted to the use of spreadsheets and/or simulation. Due to the complexity of the problem, only a small subset of the possible combinations of trucks and loaders may be considered for selection using these methods. The application of accurate mathematical modelling and cutting-edge optimisation techniques, where the optimisation is done over all possible truck and loader combinations, clearly leads to better cost savings whilst ensuring effective choices of equipment.

Our objective in this book is to present a comprehensive account of the mathematical based computational models that have been developed for determining the optimal truck–loader selection strategy for use in a large and complex mining operation. Our models not only give the optimal selection of trucks and loaders but also give the optimal allocation of the trucks and loaders. This book is organised into the following two parts: *Background and Methodology*; *Optimisation Models and Case Studies*.

The first part, *Background and Methodology*, consists of four chapters. Chapter 1 defines the equipment selection problem in surface mining and presents an introduction and relevant background to the area including some basic concepts. Chapter 2 provides a brief review of the methodology that is used in the mining industry for determining truck cycle times, equipment costs and various productivity measures for trucks and loaders. An important productivity measure is that of match factor which was first defined half a century ago. Of the available optimisation models, linear and integer programming models are the most capable for capturing the decision variables and comprehensively describing the complex relationships that exist between the various factors that arise in the equipment selection and more generally in engineering asset management equipment systems. In addition, these models are capable of handling the big and complex data sets that arise in real mines. The chapter also gives a brief overview of linear and integer programming. Chapter 3 gives a detailed literature review of the equipment selection problem in surface mining as well as the closely related equipment selection problem for the construction industry. A number of related problems are also discussed such as network design, hub location, scheduling and allocation. Models and solution procedures are reviewed. As mentioned above, match factor is an important productivity measure. Prior to our work, this measure was restricted to homogeneous fleets, and thus, applications with heterogeneous fleets were not addressed. In Chap. 4, we present our work which extends the match factor concept to more general fleets and provides an effective equipment performance measure.

The second part of our book, *Optimisation Models and Case Studies*, consists of six chapters. These chapters detail the bulk of our research on equipment selection in mining that we have carried out over a number of years. We begin by detailing the case studies that we will use in the models developed in the subsequent chapters. The case studies were provided by our industry partner. Our focus is on two case studies. The first is a simple mine with a few mining locations and 9 periods (each having one-year duration). The second case study is of a more complex mining operation having many locations with 13 periods (each having one year duration). This case study had pre-existing equipment, and this is the first time such equipment has been considered. All data are presented. In developing our mixed-integer linear programming (MILP) models, we start, in Chap. 6, by considering the simple case study of a mine having a single location and a single truck route. The objective is to determine a purchase and salvage policy for trucks and loaders that minimises the cost materials handling over a multiple period schedule. The resulting model is tested on industry data and proven to be very effective. In Chap. 7, we develop an effective MILP model for a more complex mining operation that has multiple locations and multiple periods. Pre-existing equipment and heterogeneous fleets are catered for. In addition to providing the equipment selection policy over the life of the mine, our model also gives the optimal equipment allocation. Our models are tested on two case studies. As these are large applications, we developed a pre-processing procedure and a separation algorithm to improve the tractability. These tests establish the effectiveness of our model. In our work, we accounted for equipment cost through utilised hours. Unfortunately, in real operations, equipment is not always utilised to full capacity and so the cost depends on the age of the equipment, whilst the utilisation of equipment is usually based on equipment cost. This codependency of age and utilisation is an issue and can lead to inferior solutions. In Chap. 8, we consider this issue and present a MILP model that accounts for equipment utilisation for a single location multi-period mine. Our model is successfully tested on our real case studies. We extend these notions in Chap. 9 by presenting a method for determining the cost of the equipment that accounts for utilisation. This forms the first attempt at addressing this important and difficult problem. We conclude our book with a discussion of future research directions in Chap. 10.

The bulk of the material in this book was developed over a number of years, whilst we were engaged in a Research and Development project with RioTinto. We gratefully acknowledge their support and in particular the enormous help of our collaborators Palitha Welgama and Leon Fouché. The contents of many of the chapters are from joint research publications.

Parkville, Australia
Bentley, Australia
January 2018

Christina N. Burt
Louis Caccetta

The original version of the book was revised: Originator type has been changed and the chapter author names have been included. The erratum to the book is available at https://doi.org/10.1007/978-3-319-76255-5_11

Contents

Part I Background and Methodology

1 Introduction 3
Christina N. Burt and Louis Caccetta
References 8

2 Methodology: Preliminaries and Background 11
Christina N. Burt and Louis Caccetta
2.1 Introduction 11
2.2 Truck Cycle Time 12
2.3 Shovel-Truck Productivity 13
2.4 Match Factor 14
2.5 Equipment Cost 16
2.6 Linear and Integer Optimisation 18
2.6.1 Lagrangian Relaxation 20
2.6.2 Branch and Bound 21
2.6.3 Branch and Cut 22
References 22

3 Literature Review 25
Christina N. Burt and Louis Caccetta
3.1 Introduction 25
3.2 Related Problems 26
3.3 Modelling and Solution Approaches 34
3.4 Conclusion 44
References 45

4 Match Factor Extensions 53
Christina N. Burt and Louis Caccetta
4.1 Introduction 53
4.2 Heterogeneous Truck Fleets 54

4.3 Heterogeneous Loader Fleets . . . 56
4.3.1 Example . . . 57
4.4 Heterogeneous Truck and Loader Fleets . . . 58
4.4.1 Example . . . 60
4.5 Conclusion . . . 60
References . . . 61

Part II Optimisation Models and Case Studies

5 Case Studies . . . 65
Christina N. Burt and Louis Caccetta
5.1 Introduction . . . 65
5.2 Few-Locations Case Study . . . 66
5.2.1 Locations and Routes . . . 66
5.2.2 Production Requirements . . . 66
5.2.3 Case Specific Parameters . . . 67
5.3 Many-Locations Case Study . . . 69
5.3.1 Locations and Routes . . . 69
5.3.2 Production Requirements . . . 70
5.3.3 Pre-existing Equipment . . . 70
5.3.4 Case Specific Parameters . . . 72
5.4 Compatibility and Availability . . . 74
References . . . 74

6 Single Location Equipment Selection . . . 75
Christina N. Burt, Louis Caccetta, Palitha Welgama and Leon Fouché
6.1 Introduction . . . 75
6.2 The Model . . . 76
6.2.1 Assumptions . . . 77
6.2.2 Decision Variables and Notation . . . 77
6.2.3 Objective Function . . . 78
6.2.4 Constraints . . . 81
6.2.5 Complete Model . . . 85
6.3 Computational Study . . . 85
6.4 Conclusion . . . 88
References . . . 90

7 Multiple Locations Equipment Selection . . . 91
Christina N. Burt, Louis Caccetta, Leon Fouché and Palitha Welgama
7.1 Introduction . . . 91
7.2 The Model . . . 94
7.2.1 Assumptions . . . 94
7.2.2 Decision Variables and Notation . . . 95
7.2.3 Objective Function . . . 97

7.2.4 Constraints 99
7.2.5 Complete Model 102
7.3 Computational Study 103
7.3.1 Few Locations Case Study Results 103
7.3.2 Many Locations Case Study Results 107
7.3.3 Discussion 109
7.4 Conclusion 113
References 114

8 Utilisation-Based Equipment Selection 115
Christina N. Burt, Louis Caccetta and Yao-ban Chan
8.1 Introduction 115
8.2 The Model 117
8.2.1 Assumptions 117
8.2.2 Decision Variables 118
8.2.3 Objective Function 119
8.2.4 Constraints 121
8.2.5 Complete Model 126
8.3 Validation Test Case 127
8.4 Computational Study 130
8.5 Sensitivity Analysis 134
8.6 Conclusion 141
Reference 143

9 Accurate Costing of Mining Equipment 145
Christina N. Burt and Yao-ban Chan
9.1 Introduction 145
9.2 Accurate Costing in a Non-utilisation Model 147
9.3 Utilisation and Cost Brackets in a Linear Model 148
9.4 Accurate Costing in a Utilisation Model 149
9.5 Accurate Utilisation in a Utilisation Model 151
References 152

10 Future Research Directions 153
Christina N. Burt and Louis Caccetta

Erratum to: Equipment Selection for Mining: With Case Studies E1
Christina N. Burt and Louis Caccetta

Part I
Background and Methodology

Chapter 1
Introduction

Christina N. Burt and Louis Caccetta

The ultimate goal of a mining operation is to provide a raw material at the least expense. The aspect of the mining operation which has the most influence on profit is the cost of materials handling. In this book, we focus on the problem of equipment selection for surface mines as an important driver for the overall cost of materials handling in a surface mining operation. In the mathematical branch of *Operations Research*, we interpret this problem in the context of an optimisation goal:

To optimise the materials handling such that the desired production is achieved and the overall cost is minimised.

In general, the equipment selection problem involves purchasing suitable equipment to perform a known task. It is essential that all owned equipment be compatible with both the working environment and the other operating equipment types. This equipment must also be able to satisfy production constraints even after compatibility, equipment reliability and maintenance are taken into account. By examining the equipment selection problem as an optimisation problem, we can begin to consider purchase and salvage policies over a succession of tasks or multiple periods. With this in mind, our objective is:

The original version of this chapter was revised: Originator type has been changed and the chapter author names have been included. The erratum to this chapter is available at https://doi.org/10.1007/978-3-319-76255-5_11

C. N. Burt (✉)
Department of Mathematics and Statistics, The University of Melbourne, Parkville, VIC, Australia
e-mail: c.burt@ms.unimelb.edu.au

L. Caccetta
Department of Mathematics and Statistics, Curtin University, Bentley, WA, Australia
e-mail: L.Caccetta@curtin.edu.au

C. N. Burt and L. Caccetta (eds.), *Equipment Selection for Mining: With Case Studies*, Studies in Systems, Decision and Control 150, https://doi.org/10.1007/978-3-319-76255-5_1

Given a mining schedule that must be met and a set of suitable trucks and loaders, create an equipment selection tool that generates a purchase and salvage policy such that the overall cost of materials handling is minimised.

By considering the salvage of equipment in an optimisation problem, we are effectively optimising equipment replacement as well as the selection of the equipment. Throughout the remainder of this introduction we will introduce some necessary background for the equipment selection problem and outline our approach to solving it.

The objective of the general equipment selection problem (ESP) is to choose a collection of compatible, but not necessarily homogeneous, items of equipment to perform a specified task. In many applications, the task is to move a volume of material from a set of locations to a set of destinations. However, different equipment types have attributes that can interact in a complex way with respect to productivity. In surface mining applications, the ESP addresses the selection of equipment to extract and haul mined material, including both waste and ore, over the lifetime of the mining pit. In this book, we focus specifically on the truck and loader equipment selection problem for surface mines. An important subproblem of the ESP is the equipment allocation problem, which is the problem of determining *how* and *where* the equipment should be used. This subproblem can have an enormous impact on the cost of running the equipment, as well as the resulting productivity of the mine. Therefore, in this book, we also consider the equipment allocation problem as a part of the equipment selection problem. Before we delve too deeply into problem definitions, we provide the necessary background and context to the problem.

A surface mine contains pits with mineral endowed rock (or ore). We extract ore that lies within the upper layer of the earth from surface mines [14]. This ore can include metals such as iron, copper, coal, and gold. Surface mining methods include open-pit, stripping, dredging, and mountain-top removal. This book focuses on open-pit surface mining, which involves removing ore from a large hole in the ground (sometimes referred to as a borrow-pit).

The process for creating a borrow-pit is sequential: first explosives loosen the earth; then, excavating equipment removes small vertical layers (or benches) of material (see Fig. 1.1). Over time, these benches are blasted, excavated, and removed, making the borrow-pit wider and deeper. Mining engineers categorise the mined material into ore and waste material, with subcategories that depend upon the quality

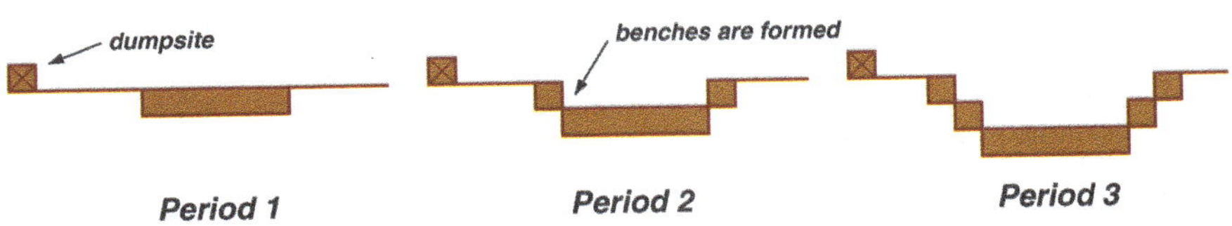

Fig. 1.1 A mining schedule is divided into planning periods. In each period, the mine planning decisions include which material will be excavated, and where the excavating equipment and stockpiles will be located. For a long-term schedule, these periods could be one year in length

or grade of the ore. Trucks transport this material to a number of dumpsites, which can include mills for crushing or refining the ore, stockpiles, and waste dumpsites. The ore is refined at the mill, while the stockpiles store supplementary material to ensure that the mill receives the correct mix of ore grades to meet market demands. The long-term mine plan optimises the timing of bench development, such that market demand is met and the value of the mine is maximised. The plan, with the optimisation of the shape of the pit, provides required productivity rates, bench sequences, and the shape of the mine (including bench heights). The height of the bench can vary from 4 to 60 meters (m) and dictates the type of equipment that can remove it. Alternate practices are available for conducting material movement in mines; however, for large-scale open-pit mining in particular, the truck-and-loader material movement practice is the preferred method of materials handling [10, 20].

Throughout this book, we consider a loader to be any type of high-productivity excavating equipment, which may include a mining loader, shovel, or excavator. Loaders lift the ore or waste material onto the trucks for removal from the mine. In an open-pit mine, loader types can include electric rope, hydraulic excavators including backhoe excavators, and front-end loaders (also called wheel loaders) [13]. Figure 1.2 illustrates these varieties, which differ significantly in terms of:

- Availability—the proportion of time the equipment is available to work [16];
- Maintenance needs—the proportion of time required for general maintenance, overhauls, and unexpected maintenance [21];
- Compatibility with different truck types—the suitability of the loader to truck height and the loader bucket to truck tray size [18];
- Volume capacity [7]; and,
- Cost per unit of production [3].

These characteristics affect the overall possible utilisation of the loading equipment and of the trucking fleet.

The type of loader selected for use in a surface mine depends on the type of mineral to be extracted and specifications of the environment, such as the bench height. We must also consider other factors in the equipment selection process, particularly the compatibility of the loaders with selected truck fleets. For example, some loaders cannot reach the top of the tray on the larger trucks. Conversely, some loader capacities exceed the capacity of the truck. If we are determined to find the best

Fig. 1.2 Excavating equipment may include (**a**) hydraulic, (**b**) rope, and (**c**) front-end loaders (images from [17])

truck and loader set, then we must model the problem such that we simultaneously select the truck and loader types.

Mining trucks, also called haul trucks or off-road trucks, haul the ore or waste material from the loader to a dumpsite. More commonly, these vary from 36 tons to 315 tons, but can be much larger. The size and cost of operating mining trucks is directly proportional to its tray capacity, while the speed at which the truck can travel is inversely proportional to its capacity. As with loaders, the variety of truck types differs according to their reliability, maintenance requirements, productivity, and operating cost.

The mine environment greatly affects the performance of a truck. For example, rimpull, which is the natural resistance of the ground to the torque of the tyre and is equal to the product of the torque of the wheel axle and the wheel radius, affects the truck's forward motion. Manufacturers supply precalculated rimpull curves for their trucks to enable a satisfactory calculation of truck cycle times. The rimpull curves map the increase in road resistance as the truck increases speed [7].

In addition, the softness of the road soil creates an effect of rolling resistance (against the truck tyres) that reduces the efficiency of the truck in propelling itself forward. Rolling resistance, which varies significantly across the road and over time, is notoriously difficult to estimate [11]. Watering and compressing the roads regularly can control and reduce the effects of rolling resistance. Haul grade, which is the incline of the haul road, can exacerbate the effects of rolling resistance and rimpull. These parameters, in addition to distance traveled, are crucial for accurately calculating truck cycle time [24]. We define truck cycle time in Chap. 2.

Loading locations include any part of the mine that provides a source of material, such as pit locations where primary excavations occur and stockpiles where reserve material is temporarily stored. Destinations include any site where material can be dumped, such as material processing locations (including crushers), stockpiles, and waste dumpsites. However, multiple origins, destinations, or pits often occur in the mine design, and the complication here is that equipment (particularly trucks) may work on any of the adjoining routes of these locations. Because several loading locations with different loading requirements may be available, different loader types may be required. The selected trucking fleet must be compatible with the loaders assigned in each period. This issue of compatibility is a complicating characteristic of surface mining equipment selection, because the trucking fleets may switch task assignments from period to period. Additionally, a partial fleet may exist at the time of equipment purchase, and because of supersession of particular models since the partial fleet was purchased (as in [8]) or some optimisation criteria, this may also lead to mixed-type (i.e., heterogeneous) fleets.

As a result of improved efficiencies after maintenance and overhauls, the operating costs of the equipment are nonlinear functions of the age of the equipment (or equipment utilisation) [5]; see Fig. 1.3. The productivity of equipment also changes over time, usually because of maintenance, equipment overhauls, operating fleet size, and driver competence. The costs are uncertain [23] because they typically encapsulate uncertain interest rates [22], depreciation [4], and revenue [2]. The presence of uncertainty makes the overall problem more difficult and can lead to infeasi-

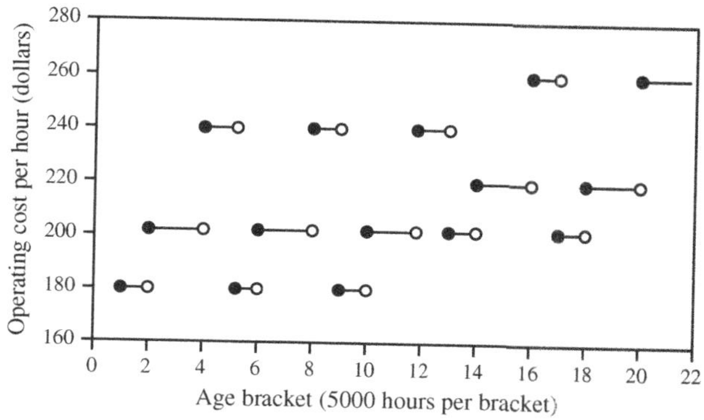

Fig. 1.3 This figure describes the discretized operating cost function over time. The rise in operating cost reflects the increased maintenance expense; large cost decreases occur when significant maintenance, such as overhaul, has taken place

bility of implemented policies. Uncertain inputs include truck cycle time [9, 20], equipment availability, truck bunching (which we describe below), and truckload variability [19].

In the context of surface mining, a robust selection of equipment can perform the required tasks on time, without compromising the mine planning. That is, we require a sufficient quantity of equipment to maintain expected productivity rates even when truck cycle times are long, some equipment is down for maintenance, or an unplanned event has taken place. Because the cost of purchasing and operating mining equipment is so high—anecdotally between 40 and 60 percent of the overall cost of materials handling [1]—robust equipment selection is a driving factor for the profitability of mining operations.

Mine planners subdivide the long-term plan (or mining schedule) into planning periods (see Fig. 1.1). The length of these periods may differ depending on the planning task: typically a year for mine scheduling decisions [15], more frequently for fleet scheduling decisions, and less frequently for equipment purchasing decisions. The mine plan dictates both the timing and manner of material movement over the strategic time horizon. Mining companies can consider long schedules (e.g., up to 50 years) in strategic planning of this nature [12]. In this time frame, replacement equipment may include types other than the original selection as a reflection of emergent technologies and changes in the mining requirements. Typically, equipment reaches replacement age after approximately 5 years for trucks and 10–15 years for loaders (depending on the type and usage). The trucks may be selected from a pool of 5–25 types [5, 21], whereas loaders could be chosen from a larger pool (e.g., 26 loader types) [4] as a result of the different variants, including rope and hydraulic, back-hoe, and front-end loaders.

The inputs to the ESP are generally (1) a long-term mining schedule, including production requirements at a number of loading and dumping locations; (2) a set of loader and truck types that may be purchased; (3) information on equipment productivity and on how it changes when equipment operates with different types of equipment; and (4) cost information, including interest and depreciation rates, purchase, maintenance, and operating costs. The output from an ESP is a purchasing strategy or policy, and ancillary information, such as how the equipment should be used with respect to defined tasks. A specific example of such ancillary information is a job allocation schedule for equipment over the defined period. Note that the allocation problem can also be solved as a subproblem of the ESP. We now formally define the ESP.

Equipment Selection Problem (Mining):
Consider the set of all truck and loader purchase policies that are feasible with respect to period demand, productivity balancing requirements between trucks and loaders, and compatibility constraints (with the environment and between equipment types). Then, the Equipment Selection Problem (ESP) is to select the minimum-cost policy from this feasible set.

Ideally, this problem would be solved in combination with the Equipment Allocation Problem, because the way that the equipment is used has an enormous impact on the cost and 'age' of the equipment.

The ESP can be solved during strategic planning, in which case the input is a long-term mine plan, or later during mining operations when new equipment is required. In the latter case, medium-term production schedules may be used as input instead of the larger resolution long-term plan. In either case, the cost of operating the equipment depends on the tasks the equipment must perform. A dimensionality difficulty lies in tying the strategic and tactical decisions of equipment types and numbers, and time of purchase, to the operational scheduling decisions over a long-term mining schedule. This disparity in time scale between strategic, tactical, and operational decisions has a noticeable effect on the effectiveness of a chosen modeling and solution approach.

Acknowledgements Components of this chapter have been published in [4, 6].

References

1. S. Alarie, M. Gamache, Overview of solution strategies used in truck dispatching systems for open pit mines. Int. J. Surf. Min. Reclam. Environ. **16**(1), 59–76 (2002)
2. P. Bodon, C. Fricke, T. Sandeman, C. Stanford, Modeling the mining supply chain from mine to port: A combined optimization and simulation approach. J. Min. Sci. **47**(2) (2011)
3. A. Bozorgebrahimi, R. Hall, M. Morin, Equipment size effects on open pit mining performance. Int. J. Surf. Min. Reclam. Environ. **19**(1), 41–56 (2005)

4. C. Burt, *An Optimisation Approach to Materials Handling in Surface Mines*. Ph.D. thesis, Mathematics and Statistics, Curtin University of Technology (Bentley, Perth, Australia, 2008), http://espace.library.curtin.edu.au:80/R?func=dbin-jump-full&local_base=gen01-era02&object_id=165737
5. C. Burt, L. Caccetta, P. Welgama, L. Fouché, Equipment selection with heterogeneous fleets for multiple period schedules. J. Oper. Res. Soc. **62**, 1498–1509 (2011)
6. C.N. Burt, L. Caccetta, Equipment selection for surface mining: A review. Interfaces **44**(2), 143–162 (2014)
7. Caterpillar Performance Caterpillar, *Handbook*, 34th edn. (Caterpillar Inc., Peoria, 2003)
8. T. Cebesoy, Surface mining equipment cost analysis with a developed linear break even model. Int. J. Surf. Min. Reclam. Environ. **11**, 53–58 (1997)
9. T. Cebesoy, M. Gözen, S. Yahşi, A systematic decision making model for optimum surface mining equipment selection, in *Mine Planning and Equipment Selection 1995: Proceedings of the Fourth International Symposium on Mine Planning and Equipment Selection, Calgary, Canada, 31st October–3 November 1995*, ed. by R.K. Singhal, A.K. Mehrotra, J. Hadjigeorgiou, R. Poulin (A.A. Balkema, Rotterdam, 1995), pp. 369–377
10. J.M. Czaplicki, On a number of spare haulage units for the shovel-truck system in open pit mines. Int. J. Surf. Min. Reclam. Environ. **6**, 57–59 (1992)
11. P.S. Dunston, J.V. Sinfield, T.-Y. Lee, Technology development decision economics for real-time rolling resistance monitoring of haul roads. J. Constr. Eng. Manag. **133**(5), 393–402 (2007)
12. R. Epstein, S. Gaete, F. Caro, A. Weintraub, P. Santibanez, J. Catalan, Optimizing long term planning for underground copper mines, in *Proceedings of the Copper 2003-Cobre 2003, 5th International Conference*, vol. I (CIM and the Chilean Institute of Mining, Santiago, 2003)
13. S. Erçelebi, C. Kirmanli, Review of surface mining equipment selection techniques, in *Mine Planning and Equipment Selection 2000: Proceedings of the Ninth International Symposium on Mine Planning and Equipment Selection /Athens/Greece/6–9 November 2000*, ed. by G.N. Panagiotou, T.N. Michalakopoulos (A.A. Balkema, Rotterdam, 2000), pp. 547–553
14. C. Fricke, *Applications of Integer Programming in Open Pit Mining*. Ph.D. thesis, Department of Mathematics and Statistics, University of Melbourne (2006)
15. A. Gleixner, Solving large-scale open pit mining production scheduling problems by integer programming. Master's thesis, Technische Universität Berlin (2008)
16. R.A. Hall, L.K. Daneshmend, Reliability modelling of surface mining equipment: Data gathering and analysis methodologies. Int. J. Surf. Min. Reclam. Environ. **17**(3), 139–155 (2003)
17. Immersive Technologies (2012), http://www.immersivetechnologies.com/
18. B. Morgan, Optimizing truck-loader matching, in *Mine Planning and Equipment Selection 1994: Proceedings of the Third International Symposium on Mine Planning and Equipment Selection, Istanbul, Turkey, 18–20 October 1994*, ed. by A.G. Pasamehmetoglu (A.A. Balkema, Rotterdam, 1994), pp. 313–320
19. A. Newman, E. Rubio, R. Caro, A. Weintraub, K. Eurek, A review of operations research in mine planning. Interfaces **40**(3), 222–245 (2010)
20. C. Ta, J. Kresta, J. Forbes, H. Marquez, A stochastic optimization approach to mine truck allocation. Int. J. Surf. Min. Reclam. Environ. **19**(3), 162–175 (2005)
21. E. Topal, S. Ramazan, A new MIP model for mine equipment scheduling by minimizing maintenance cost. Eur. J. Oper. Res. **207**(2), 1065–1071 (2010)
22. W. Wiesemann, D. Kuhn, B. Rustem, Maximizing the net present value of a project under uncertainty. Eur. J. Oper. Res. **202**(2), 356–367 (2010)
23. Y. Zhang, *Advances in LTL load plan design*. Ph.D. thesis, Georgia Institute of Technology (2010)
24. L. Zhongzhou, L. Qining, Erlangian cyclic queueing model for shovel-truck haulage systems, in *Mine Planning and Equipment Selection*, ed. by R. Singhal (1988), pp. 423–428

Chapter 2
Methodology: Preliminaries and Background

Christina N. Burt and Louis Caccetta

2.1 Introduction

In this chapter, we provide a broad introduction into the various performance measures and strategies for obtaining solutions for the equipment selection problem. There are many ways to broach a problem—each with its own advantages and focus. Heuristic or approximate solution approaches, for example, might be computationally efficient, while exact approaches might bring a higher degree of accuracy and optimality to a solution that equates to savings or productivity improvements. In order to prepare the reader for the literature chapter (Chap. 3), in this chapter we have provided a background on the key strategies that appear in the literature. The truck cycle time features in all solution strategies for the equipment selection problem. Therefore, we provide a definition of truck cycle time, along with explanations of how it is typically calculated. We follow this with an introductory description of shovel-truck productivity and match-factor—two performance measures that have been used extensively for their simplicity and ease of application. However, these approaches do not account for the cost of the equipment. In order to achieve a cost model, we first describe the ways that cost can be evaluated. We then describe some

---

The original version of this chapter was revised: Originator type has been changed and the chapter author names have been included. The erratum to this chapter is available at https://doi.org/10.1007/978-3-319-76255-5_11

C. N. Burt (✉)
Department of Mathematics and Statistics, The University of Melbourne, Parkville, VIC, Australia
e-mail: c.burt@ms.unimelb.edu.au

L. Caccetta
Department of Mathematics and Statistics, Curtin University, Bentley, WA, Australia
e-mail: L.Caccetta@curtin.edu.au

C. N. Burt and L. Caccetta (eds.), *Equipment Selection for Mining: With Case Studies*, Studies in Systems, Decision and Control 150,
https://doi.org/10.1007/978-3-319-76255-5_2

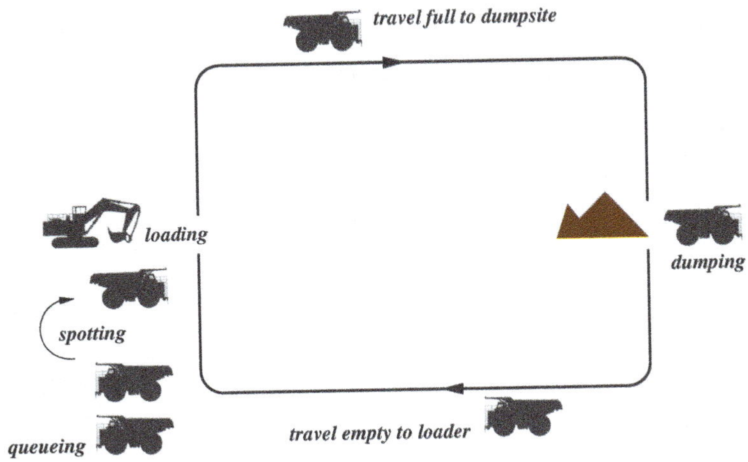

Fig. 2.1 The truck cycle time is measured from the time the truck is filled at the loader, travels full to the dumpsite, dumps the load, and travels empty to the loader to join a queue and position itself for the next load (spotting). The truck cycle time includes queuing and waiting times at the dumpsite and loader (images adapted from [16])

methodology that accounts for cost, focussing on the optimisation approaches of linear programming, integer programming and mixed-integer programming.

2.2 Truck Cycle Time

Definition 2.2.1 The **truck cycle time** comprises of load time, haul time (full), dump time, return time (empty), queuing and spotting [Fig. 2.1].

A cycle may begin at a loader where the truck receives its load. The truck then **travels full** to the dump-site via a designated route along a haul road. The dump-site may be a stockpile, dump-site or mill. Once the load has been dumped, the truck turns around and **travels empty** back to the loader where it joins a queue of trucks to be loaded. The act of manoeuvring the truck under the loader to be served is called **spotting**. This can take several minutes. In a large mine the truck cycle time may be around 20–30 minutes in total, and can vary a lot over time as the stockpiles are moved and the mine deepens. The cycle time may also vary if the truck returns to a different loader with alternate specifications, or returns to a different loading location. These variants will be considered in this book.

Ultimately we wish to include low-level details of the mine, such as topography and rolling resistance, in the modelling process. These parameters can be estimated prior to modelling and incorporated into the truck cycle time. Further, the truck cycle time can be used to absorb parameters such as rimpull, haul grade and haul distance into one estimate. However, the level of queuing that occurs in a fleet is dependent on the number of trucks operating against each loader. This makes it difficult to accurately estimate truck cycle times before the fleet is determined.

In industry, the common method of truck cycle time estimation is to estimate the speed of the trucks using manufacturers' performance guidelines [25]. These guidelines are simulation results that take into consideration engine power, engine transmission efficiency, truck weight, capacity, rimpull, and road gradients and conditions [3]. This is combined with topographical information to provide an estimate of the hauling route. The guidelines must also be used in combination with rolling resistance estimates to determine any lag in cycle time. Smith et. al. [25] provides a method for determining a rolling resistance estimate. Regression models can also be used to determine good truck cycle time estimates [7]. In this book, we make use of truck cycle time estimates provided by an industry partner.

2.3 Shovel-Truck Productivity

The ability to predict the productivity of a truck and loader fleet is an important problem for mining and construction, as the productivity of the fleet is intrinsic to equipment selection. A part of the equipment selection liturature bases the selection entirely on productivity estimates of the fleets. This research usually comes under the banner of *shovel-truck productivity*, and focuses on "predicting the travel times on the haul and return portions of the truck cycle … and the prediction of the interaction effect between the shovel and truck at the loading point" [21]. The shovel-truck productivity problem has been well established in the construction and earthmoving literature [17]. This work aims to match the equipment (in both type and fleet size) such that the productivity of the overall fleet is maximised. However, much of the literature on shovel-truck productivity exists for construction case studies and little published research applies to surface mining. Nonetheless these methods must be addressed here as they represent the core ideas behind current industry practice in surface mining equipment selection [12, 25].

Those methods deemed classical include **match factor** and **bunching theory**. The match factor is the ratio of truck arrival rate to loader service time, and provides an indication of the efficiency of the fleet. Bunching theory studies the natural variance in the truck cycle time due to bunching of faster trucks behind slower trucks. Shovel-truck productivity methods incorporate both match factor and bunching ideas into the solution. These methods use many assumptions, considerable expert knowledge/experience and rely on heuristic solution methods to achieve a solution. Modelling of the true bunching effect would be a helpful asset to the mining and construction industries, as the effect is not well studied and is currently unresolved. However, the derivation of such a model is beyond the scope of this book.

2.4 Match Factor

The match factor itself provides a measure of productivity of the fleet. The ratio is so called because it can be used to match the **truck arrival rate** to **loader service rate**. This ratio removes itself from equipment capacities, and in this sense, potential productivity, by also including the loading times in the truck cycle times.

Douglas [10] published a formula that determined a suitable number of trucks, M_b, to *balance* loader output. This formula is the ratio of loader productivity to truck productivity, but as it makes use of equipment capacity it is considering the *potential* productivity of the equipment. That is, if the loader is potentially twice as productive as the selected truck type, then we require two trucks to balance the productivity level. Let c_e denote the capacity of equipment type $e \in \mathbf{X} \cup \mathbf{X}'$, and t_e signify the cycle time of equipment type e, where $\mathbf{X}$ is the set of all truck types and $\mathbf{X}'$ is the set of all loader types. The productivity of equipment type e is represented by P_e and the number of trucks of type i in the fleet is x_i, where $i \in \mathbf{X}$, while we denote the loader types as i' where $i' \in \mathbf{X}'$. We denote the equipment efficiency by E_e (representing the proportion of time that the equipment is actually producing). We can write

$$P_{i'} = \frac{c_{i'} E_{i'}}{t_{i'}} \qquad \forall \ \ i' \in \mathbf{X}', \tag{2.1}$$

for a single loader operation. The productivity of the truck fleet is represented by:

$$P_i = \frac{c_i E_i x_i}{t_i} \qquad \forall \ \ i \in \mathbf{X}, \tag{2.2}$$

and the match balance is represented by:

$$M_b = \frac{P_{i'}}{P_i}. \tag{2.3}$$

Truck cycle time is defined for Eq. (2.2) as the sum of non-delayed transit times, and includes haul, dump and return times. Note that ratio (2.3) is restricted to one loader. This is a simple ratio that can be used to ensure that the truck and loader fleets do not restrict each other's capacity capabilities. Sometimes however, it is not necessary for the productivities of the truck and loader fleets to be perfectly matched. Morgan and Peterson [21] published a simpler version of the ratio, naming it the match factor, $MF_{i,i'}$, for truck type i working with loader type i' is given as:

$$MF_{i,i'} = \frac{t_{i,i'} x_i}{\bar{t}_X y_{i'}}, \tag{2.4}$$

where x_i is the number of trucks of type i; $y_{i'}$ is the number of loaders of type i'; $t_{i,i'}$ is the time taken to load truck type i with loader type i'; and $\bar{t}_X$ is the average cycle time for the trucks excluding waiting times. This ratio uses the actual productivities of the equipment rather than potential productivities, and therefore achieves a different result to Eq. (2.3). In this book, we consider only the Morgan and Peterson [21] interpretation of match factor: we are interested in the actual productivities of the truck and loader fleets.

Definition 2.1 The **match factor** is the ratio of actual truck arrival rate to loader service time.

In this book, we make use of the match factor as a productivity indicator, and contrary to the Morgan and Peterson [21] interpretation, we assume that queue and wait times are included in the cycle times. With this idea of cycle time in mind, a match factor of 1.0 represents a balance point, where trucks are arriving at the loader at the same rate that they are being served. Typically, if the ratio exceeds 1.0 this indicates that trucks are arriving faster than they are being served. For example, a high match factor (such as 1.5) indicates over-trucking. In this case the loader works to 100% efficiency, while the trucks must queue to be loaded.

A ratio below 1.0 indicates that the loaders can serve faster than the trucks are arriving. In this case we expect the loaders to wait for trucks to arrive. For example, a low match factor (such as 0.5) correlates with a low overall efficiency of the fleet, namely 50%, while the truck efficiency is 100% (see Fig. 2.2). This is a case of under-trucking; the loader's efficiency is reduced while it waits. Unfortunately, in practice a theoretical match factor of 1.0 may not correlate with an actual match factor of 1.0 due to truck bunching. In this sense, the calculated match factor value is optimistic.

The match factor ratio has been used to indicate the efficiency of the truck or loader fleet and in some instances has been used to determine a suitable number of trucks for the fleet [8, 18, 24]. While the ratio can be used to give an indication of efficiency or productivity ratios, it fails to take truck bunching into account. Therefore caution must be taken in the interpretation of any calculated match factor values.

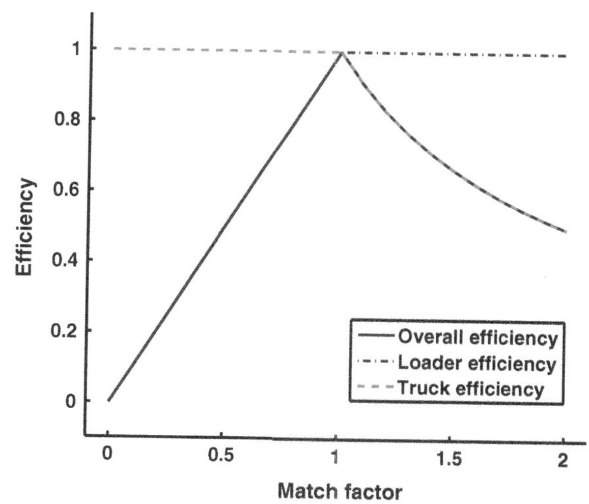

Fig. 2.2 The match factor (MF) is the ratio of loader productivity to truck productivity. A low MF (< 0.5) suggests that the loader is not working at capacity, whereas a high MF (> 1) suggests the truck fleet is smaller than necessary to maintain a productivity balance between truck and loader fleets

Match factor has been adopted in both the mining and construction industries [20, 24]. The construction industry is interested in achieving a match factor close to 1.0, which would indicate that the productivity levels of the fleet are maximised. However, the mining industry may be more interested in lower levels of match factor (which correspond to smaller trucking fleets and increased waiting times for loaders) as this may correlate with a lower operating cost for the fleet. This can happen if equipment with greater productivity rates than required can perform the task with lower operating costs than equipment that perfectly matches the required production.

The match factor ratio relies on the assumption that the operating fleets are **homogeneous**. That is, only one type of equipment for both trucks and loaders is used in the overall fleet. When used to determine the size of the truck fleet, some literature simplifies this formula further by assuming that only one loader is operating in the fleet (see [20, 22, 24]). Homogeneous fleets are desirable for the mine, as they simplify maintenance, training of artisans and the burden of carrying spare parts for different types of equipment. However, heterogeneous fleets may provide overall cost savings.

In practice, mixed fleets and multiple loaders are common due to **pre-existing** equipment or optimal fleet selection that minimises the cost of the project. A situation with pre-existing equipment can arise both at the start of a mining schedule, and when a new selection of equipment is desired part-way through the schedule. This highlights a need for a match factor ratio that can be applied to heterogeneous fleets. In Chap. 4, we will extend the match factor concept to more general fleets.

2.5 Equipment Cost

The operating cost of mining equipment dominates the overall cost of materials handling over time. Typically these costs include maintenance, repairs, tyres, spare parts, fuel, lubrication, electricity consumption and driver wages into one estimate. The best way to account for the operating cost of mining equipment is, in itself, an important problem. Some equipment selection tools use life-cycle costing techniques to obtain an **equivalent unit cost** for the equipment [4]. These costs estimate the average lifetime cost *per hour* or *per tonne*. Clearly this is not practical if we are considering salvaging equipment when it is no longer useful or has reached the end of its optimal replacement cycle. Industry improves the equivalent unit cost estimate by scaling the value depending on the age of the equipment. That is, if the equipment requires a full maintenance over-haul at the age of 25,000 hours then this cost bracket will reflect a greater expense through a scaled factor of the unit cost.

Equipment operating cost is highly dependent on the age of the equipment. That is, cost per tonne is determined by productivity. Equipment productivity is dependent on equipment availability, while equipment availability is dependent on equipment age. Operating cost can also be affected substantially by the simple addition of one loader to a single-loader fleet [2]. Although the most obvious objective function for an equipment selection model is to minimise cost, as a function of utilisation and equipment age this adds great complexity to the problem and has the potential to introduce nonlinearities [13].

Any mining operation is dynamic in nature and may be subject to considerable changes in the mine plan. In many cases, an equipment selection plan for a multi-period mine may be rendered inadequate as these changes come to light. The purpose of the tools derived in this book, however, aim to provide the best possible starting solution given the information available at the time. To add to this varying nature of the production parameter, the cost parameters may also change significantly and are themselves estimates [13]. Specifically, the capital expense data available at the time the equipment selection tool is run may differ from the time of purchase due to:

- the establishment of new contracts with the corresponding suppliers;
- improved historical data (accumulated through previously owned equipment) which may be combined with supplied data (from the equipment producers) [11, 25];
- a change in demand for second-hand equipment or scrap metal—thus affecting the salvage value of a piece of machinery;
- changes in the interest and depreciation rates used for the *net present value* calculations.

As its name suggests, the *net present value* (NPV) is the difference between money in-the-hand now and the value of that money if it has been invested for a set amount of time, at a certain interest rate. In the mining industry, NPV is a term used more broadly to capture the change in value of money over time. This is important, because in long-term scheduling (which may be planned over 50 years) we need to be able to compare the cost of a decision made now with a decision made in the future. This can be achieved quite simply by multiplying costs by the following expression:

$$\frac{1}{(1+I)^t},$$

where I is the fixed interest rate and t is the future time period (number of years in the future) for comparison. This expression is called a *discount factor*, as it discounts costs to the present to allow comparison. One limitation of this approach, from a modelling perspective, is that it is much more convenient if t is known, and not a variable in the problem. This allows a simple linear formulation of the ESP. On the other hand, if t were a variable, then this expression alone would be nonlinear and would lead to a messier formulation of the ESP.

With these examples as justification, we argue that it is not necessary to consider the cost objective function in its most natural and accurate form: nonlinear. As all the parameters of the objective function are themselves approximations, the objective function may be more wisely considered in piecewise linear format. Certainly in industry this is common practice where operating cost, for example, is considered to be a piecewise linear function of an age bracket, rather than a nonlinear function of unit age. By these arguments, the relative parameters of a linearised objective function can be considered to be sufficiently precise. Using hire cost data is a simple alternative to using a mix of manufacturer supplied production costs and real data [11], but this is not always possible or practical.

2.6 Linear and Integer Optimisation

In an optimisation problem, we focus on a single **objective function**, $f(x)$, whose purpose is to measure the quality of the decision [19]. Mathematical programs look at the state of a system and its structure, and in considering a suitable objective determines how the system can move into the next state.

A general mathematical program can be formulated as follows:

$$\min \quad f = c^T x$$

$$\text{subject to} \quad Ax \leq b, \tag{2.5}$$

$$x \geq 0, \tag{2.6}$$

$$x_j \in Z^{\geq 0} \qquad \forall j \in J \subset \{1, 2, \ldots, i\} | i \leq n. \tag{2.7}$$

The input data consists of the matrices c $(1 \times n)$, A $(m \times n)$ and b $(m \times 1)$, and the n vector, x, of variables whose values are to be determined. Note we have an LP when $i = 0$ and a pure IP when $i = n$. If the objective function $f = c^T x^2$, then we have a quadratic program (QP).

Linear programming is a mathematical programming technique that aims to capture the behaviour of the problem within a linear objective function and linear constraints. This technique is credited with both explicit formulation of the problem and, through various solution methods, an efficient solution. The philosophy of linear programming is simply to derive a mathematical structure by observing the important components of the system and their essential interrelationships [9]. The "Transportation Problem" is a famous example of linear programming.

For **integer programming**, we have the additional restriction that all variables are integers. The appeal of integer programming as a modelling method is the compactness of model presentation, the existence of proof of optimality for many of the solution methods (such as branch and bound), and the ability to perform sensitivity analysis on the objective function and constraints after solving. However, **mixed integer linear programs** (including integer programs with some binary 0–1 variables) are at times computationally difficult. Some aspects of the formulation have an enormous impact on the computation time, such as the integrality and formation of the constraint matrix [26].

Mathematical modelling can bring more advantages in analysis than simply the concise and comprehensible structuring of the problem. The way in which a problem is modelled can help to identify "cause-and-effect relationships" [15]. Further to this, the various relationships between the variables are considered simultaneously.

The theory and application of linear and integer programming has undergone considerable development and advancement since the early 1950s. This is clearly evident from the vast literature that has accumulated over the ensuing decades and the current trend of increasing activity in this area. This growth, particularly in integer programming, has been greatly accelerated by advances in computer technology. These advances have facilitated the development of sophisticated computational mathematical techniques for solving the many complex problems arising in modern business and industry. Further, the continued need for business and industry to efficiently utilize the limited and expensive resources to survive in the present and future highly competitive global environment will ensure that mathematical programming, in particular integer programming, remains an active area in the foreseeable future.

In many applications the problem that arises is one of optimising a function (representing profit, output or cost) subject to a specified set of constraints (representing the limited resources and/or the operational requirements of the system). Large instances of these linear optimisation problems arise in many applications including: airline crew scheduling; data association; network design and analysis; network routing; production planning; resource allocation; financial management and planning; facility layout design; design of automated systems; human resource planning; location of service facilities; and many more. In the mining industry, mathematical programming provides accurate and effective mathematical models that capture the geology and structure of the ore-body as well as the economic, the metallurgical and the geotechnical factors that are essential in mine planning and management. In addition to establishing the economic potential and viability of a mining operation, these models provide the framework in which to develop the smart computational algorithms for the design of optimum pits and the determination of optimal production schedules, optimal product blends, optimal operating layouts, effective strategies for a range of transport and logistics issues, an efficient mine site rehabilitation program and as we shall see in this book effective strategies for equipment selection and allocation. In this section we briefly introduce the important area of linear and integer programming.

Efficient and easily available commercial LP and MILP packages has greatly assisted the application of linear and integer programming methods to large scale industrial problems. In addition to providing powerful optimisation solvers the technology also provides effective algebraic modelling tools such as GAMS, AMPL and AIMMS, that allow users to express the LP/MILP problems in a natural mathematical form. Commercial solvers include CPLEX, Gurobi, Xpress and LINDO, to name just a few. In our case studies we use Ilog Concert Technology with the CPLEX solver.

We now briefly detail some basic approaches for solving MILPs. We focus on the exact methods of Lagrangian Relaxation, Branch and Bound, and Branch and Cut. As MILPs are for the most part NP-hard, many heuristic procedures have been proposed. In some cases, these search procedures produce good approximate solutions, particularly when the structure of the problem can be exploited. Fast heuristics capable of producing good approximations are important in the success of exact methods. For an excellent account of search methods in optimisation we refer to the book by [1].

2.6.1 Lagrangian Relaxation

Many MILPs can be viewed as easy problems complicated by a relatively small set of difficult constraints. The Lagrangian Relaxation method dualizes these difficult constraints and then attempts to solve the resulting relaxed problem. For example, consider the above MILP written as :

$$\min \quad f = c^T x \tag{2.8}$$

$$\text{subject to} \quad A_1 x \leq b_1, \qquad \text{(difficult constraints)} \tag{2.9}$$

$$A_2 x \leq b_2, \qquad \text{(easy constraints)} \tag{2.10}$$

$$x_j \in Z^{\geq 0} \qquad \forall j \in J \subseteq \{1, 2, \ldots, i\} | i \leq n. \tag{2.11}$$

The Lagrangian relaxation is :

$$g(\lambda) = \min\{c^T x + \lambda\,(A_1 x - b_1)\} \tag{2.12}$$

$$\text{subject to} \quad (2.10) - (2.11), \tag{2.13}$$

where λ is a non-negative vector of Lagrangian multipliers.

For a given λ the relaxed problem is easy to solve. Observe that if x^* is an optimal solution for (2.8)–(2.11), then

$$g(\lambda) \leq c^T x^* + \lambda(A_1 x^* - b_1) \leq f(x^*). \tag{2.14}$$

Thus any solution to the relaxed problem provides a lower bound in the objective function value of the original MILP. The best choice for λ is that which yields the greatest lower bound, or equivalently any λ which is optimal in the dual problem. The optimality conditions are:

Theorem 2.1 *For a given $\bar{\lambda}$, suppose $\bar{x}$ satisfies:*

1. *$\bar{x}$ is an optimal solution to* (2.12), *and* (2.10)–(2.11).
2. $A_1 \bar{x} \leq b_1$.
3. $\bar{\lambda}\,(A_1 \bar{x} - b_1) = 0$.

Then $\bar{x}$ is an optimal solution to the original MILP (2.8)–(2.11).

The solution to the dual problem may be difficult because of its discrete nature. In practice duality gaps may arise. The subgradient optimisation method [14] provides a numerical technique for reducing the duality gaps. Theoretically, it is guaranteed to determine the optimal λ. However, in practice convergence may be slow and the optimal λ may not be achievable in a realistic time. This is the major disadvantage of the method. An advantage of the method is that together with a heuristic that generates a feasible solution, it provides a measure for evaluating the quality of the best available solution through upper and lower bounds. Usually, once a feasible solution that is within a specific tolerance of the optimum is obtained, the computation ceases and the current solution is accepted. An alternative strategy for obtaining a solution is to implement a Branch and Bound method when a duality gap is encountered (branch on a fractional variable).

2.6.2 Branch and Bound

The method of branch and bound has been effectively used to solve a number of computationally difficult problems. Basically, the idea is to subdivide (branch) the feasible solution set into successively smaller subsets, placing bounds on the objective function value over each subset, and using these bounds to discard subsets from further consideration and to select the next subset to further subdivide. The branching process is carried out by fixing the value of the branching variable. In the case of a feasible problem, the process stops when we have a solution to the original problem which has an objective function value, in the case of a minimisation problem, less than or equal to all lower bounds of the generated subsets. Consider the MILP (2.8)–(2.11). A relaxed problem can be obtained by dropping the integer restrictions (2.11). So the relaxed problem is the LP (2.8)–(2.10), and $x \geq 0$. The optimal solution to the relaxed LP provides an initial lower bound for the objective function value of the original MILP. This solution is of course optimal if (2.11) are also satisfied. Observe that any feasible solution of the MILP provides a valid upper bound.

In many applications upper bounds are obtained by the application of a fast heuristic. Lower bounds can be used in a tree search technique to specify additional restrictions. Note that nodes in the tree correspond to subsets generated, the root node to the initial relaxed feasible solution set. If at any node in the search tree one has a lower bound for a subset that is greater than or equal to the current upper bound, then we do not need to consider this subset any further (i.e. no further branching from this node is done).

The branch and bound method can be implemented in many ways. The success of the implementation is highly dependent on: the branching strategy; the search strategy; and the quality of the lower and upper bounds generated. The usual search strategy is depth-first search, where a subset chosen from the list is explored until either it violates the lower bound criteria or an improved feasible solution is obtained. Branching occurs from the parent node in the branch that has not yet been fully explored. An alternative is breadth-first search where the branching is done from the subset with the lowest lower bound. Lower bounds can be obtained at each node by either solving the corresponding relaxed problem optimally or by the application of an approximation method (for example, application of subgradient optimisation to a Lagrangian relaxation problem).

2.6.3 Branch and Cut

The method of branch and cut is a powerful technique for solving MILP problems. Basically, this method attempts to strengthen the lower bounds by the addition of constraints (cuts) at each node within a branch and bound procedure. Consider the MILP (2.8)–(2.11); denote this problem by P. Let K be a set of valid inequalities for the problem P—that is, a set of inequalities that potentially improve the continuous relaxation of P, yet do not eliminate any feasible solutions from the set of all solutions. Now consider the relaxed problem is P':

$$\min \quad f = c^T x \tag{2.15}$$

$$\text{s.t.} \quad Bx \leq d \tag{2.16}$$

$$x \geq 0. \tag{2.17}$$

Here, (2.16) is a subset of the original constraints (2.9) and (2.10). Also $\{Ax \leq b\} \backslash \{Bx \leq d\} \subseteq K$. A lower bound for (2.8)–(2.11) can be generated from the following procedure adapted from [23].

begin
Step 1 : Let $L = \emptyset$.
Step 2 : Solve P with the additional constraints in L, and let $\bar{x}$ be the optimal solution.
Step 3 : Find one or more inequalities in K that are violated by $\bar{x}$.
Step 4 : If none are found, stop. Otherwise add the violating inequalities to L and go to Step 2.
end

Algorithm 1: Cutting plane procedure

The above cutting plane procedure terminates when no further violation can be found or an optimal solution has been found. The success of the above procedure is highly dependent upon efficiently finding sets of 'strong' inequalities of K. The problem of finding a violating inequality of K or proving that no such inequality exists for solving is commonly referred to as the 'separation problem'. Ideally an efficient exact method to solve the separation problem (Step 3) is required for completeness. This is often computationally difficult, so we resort to heuristic search methods [6].

Acknowledgements Part of this chapter has been published in [5] and [6].

References

1. E. Aarts, J.K. Lenstra, *Local search in combinatorial optimization* (Wiley, Chichester, 1997)
2. S. Alkass, K. El-Moslmani, M. AlHussein, A computer model for selecting equipment for earthmoving operations using queuing theory. Constr. Inform. Digit. Libr. (2003)
3. G.H. Blackwell, Estimation of large open pit haulage truck requirements. CIM Bull. **92**(1028), 143–148 (1999)

4. A. Bozorgebrahimi, R. Hall, M. Morin, Equipment size effects on open pit mining performance. Int. J. Surf. Min. Reclam. Environ. **19**(1), 41–56 (2005)
5. C. Burt, An optimisation approach to materials handling in surface mines. Ph.D. thesis, Mathematics and Statistics, Curtin University of Technology, Bentley, Australia, 2008, http://espace.library.curtin.edu.au:80/R?func=dbin-jump-full&local_base=gen01-era02&object_id=165737
6. L. Caccetta, Branch and cut methods for mixed integer linear programming problems, in *Progress in Optimization*, ed. by Y. Xiao-qi, A.I. Mees, M. Fisher, L. Jennings. Applied Optimization Series, vol. 39 (Kluwer Publishers, Dordrecht, 2000) pp. 21–44
7. N. Çelebi, An equipment selection and cost analysis system for openpit coal mines. Int. J. Surf. Min. Reclam. Environ. **12**, 181–187 (1998)
8. N. Cetin, Open-pit truck/shovel haulage system simulation. Ph.D. Thesis, Middle East Technical University, (2004), http://etd.lib.metu.edu.tr/upload/12605499/index.pdf
9. G. Dantzig, *Linear Programming and Extensions*. (Princeton University Press, New Jersey, 1998)
10. J. Douglas, Prediction of shovel-truck production: A reconciliation of computer and conventional estimates. Technical Report No. 37 (Stanford University, California, 1964)
11. D. Edwards, H. Malekzadeh, S. Yisa, A linear programming decision tool for selecting the optimum excavator. Struct. Surv. **19**(2), 113–120 (2001)
12. S. Erçelebi, C. Kirmanli, Review of surface mining equipment selection techniques, in *Mine planning and equipment selection 2000*, ed. by G.N. Panagiotou, T.N. Michalakopoulos. Proceedings of the Ninth International Symposium on Mine Planning and Equipment Selection /Athens/Greece/6–9 Nov 2000, (Rotterdam, 2000), pp. 547–553 (A.A. Balkema)
13. M. Hassan, G. Hogg, D. Smith, A construction algorithm for the selection and assignment of materials handling equipment. Int. J. Prod. Res. **23**(2), 381–392 (1985)
14. M. Held, P. Wolfe, H.P Crowder, Validation of subgradient optimization. Math. Program. **6**(1), 62–88 (1974)
15. F. Hillier, G. Lieberman, *Introduction to Operations Research*, 5th edn. (McGraw-Hill Publishing Company, New York, 1990)
16. Immersive Technologies (2012), http://www.immersivetechnologies.com/
17. A. Kesimal, Shovel-truck productivity and efficiency studies for overburden removal in an open-pit coal mine. Trans. Inst. Min. Metall. Sect. A: Min. Ind. **107**, A37–A40 (1998)
18. Y. Kuo, Highway earthwork and pavement production rates for construction time estimation. Ph.D. Thesis, University of Texas, (2004), http://dspace.lib.utexas.edu/bitstream/2152/208/1/kuoy81347.pdf
19. D. Luenberger, *Linear and Nonlinear Programming*, 2nd edn. (Kluwer Academic, Boston, 2003)
20. B. Morgan, Optimizing truck-loader matching, in *Mine Planning and Equipment Selection 1994*, ed. by A.G. Pasamehmetoglu. Proceedings of the Third International Symposium on Mine Planning and Equipment Selection, Istanbul, Turkey, 18–20 Oct 1994, (Rotterdam, 1994), pp. 313–320. (A.A. Balkema)
21. W.C. Morgan, L.L. Peterson, Determining shovel-truck productivity. Min. Eng. 76–80 (1968)
22. S.W. Nunnally, *Managing Construction Equipment*, 2nd edn. (Prentice-Hall Inc., New Jersey, 2000)
23. M. Padberg, G. Rinaldi, A branch-and-cut algorithm for the resolution of large-scale symmetric traveling salesman problems. SIAM Rev. **33**(1), 60–100 (1991)
24. S.D. Smith, J.R. Osborne, M.C. Forde, Productivity estimation in back-acter/dump-truck earthmoving operations. Proc. Inst. Civ. Eng. Transp. **111**, 125–131 (1995)
25. S.D. Smith, G.S. Wood, M. Gould, A new earthworks estimating methodology. Constr. Manag. Econ. **18**(2), 219–228 (2000)
26. H.A. Taha, *Integer Programming: Theory, Applications, and Computations* (Academic Press, New York, 1975)

Chapter 3
Literature Review

Christina N. Burt and Louis Caccetta

3.1 Introduction

The literature discusses two approaches to solving the equipment selection problem: (1) to partition the problem and solve each partition sequentially; and (2) to develop holistic computational models. The most common approach in the mining equipment selection literature has been to take a sequential approach (e.g., first select loader type, then select truck type, and finally determine fleet sizes). However, by observing recent advancements in related research in mathematical programming, especially in applications with a similar problem structure, the mining industry may be able to solve larger-scale, more difficult instantiations of the problem. In particular, these advancements could lead mine planning away from sequential decision making for problems that are essentially interdependent and should therefore be solved holistically.

In later Chaps. (6, 7 and 8), we provide examples of mixed-integer programming formulations of the ESP that illustrate the following description of the general structure of the problem. A fixed-charge objective function results from considering purchase, salvage, and service decisions in a cost-minimisation scenario. A fixed

---

The original version of this chapter was revised: Originator type has been changed and the chapter author names have been included. The erratum to this chapter is available at https://doi.org/10.1007/978-3-319-76255-5_11

C. N. Burt (✉)
Department of Mathematics and Statistics, The University of Melbourne, Parkville, VIC, Australia
e-mail: c.burt@ms.unimelb.edu.au

L. Caccetta
Department of Mathematics and Statistics, Curtin University, Bentley, WA, Australia
e-mail: L.Caccetta@curtin.edu.au

C. N. Burt and L. Caccetta (eds.), *Equipment Selection for Mining: With Case Studies*, Studies in Systems, Decision and Control 150,
https://doi.org/10.1007/978-3-319-76255-5_3

charge represents an incremental (disjoint) jump in the objective function and is usually the result of purchase or other binary decisions. However, the capacity of the loaders and trucks has limitations; the productivity of the loaders can depend on the pairing with particular truck types, and the productivity of both trucks and loaders is influenced by the number of bucket loads required to fill the tray. These factors, in combination with the multiple-flow paths the equipment may take in transit, results in a problem with a structure similar to the fixed-charge capacitated multicommodity flow problem (e.g. see [63]). However, the underlying transportation network and arising transportation problem is often very simple, and is composed of a small number of excavation and dumping locations, often with some shared routes.

In the construction industry, truck and loader equipment selection is very similar to surface mining equipment selection. The principle problem involves selecting appropriate loading and trucking equipment. However, the transportation networks are very simple. Beyond this, a key difference lies in the objective of the operations—in the construction industry, the earliest finish time of the project (i.e., shortest makespan) is very important. The objective function is not the only difference: the scale of material moved is significantly smaller in construction operations compared to mining operations. Two other similar applications, with respect to problem structure, are manufacturing production research (including equipment selection and allocation problems) and capacitated network design (in the presence of multicommodity flow). To help apply theoretical advancements that are of practical use to the surface mining community, we include literature from these applications in this chapter where it is appropriate. However, our primary focus is the mining and construction literature.

We first outline some related problems. We then provide a review of modelling and solution methods in both the mining and operations research (OR) literature. Using the OR literature as a guide, we conclude with future directions for research in the context of surface mining applications.

3.2 Related Problems

In this section, we outline problems in mining that are similar to the ESP. We also provide a summary of the relevant papers in Tables 3.1–3.2. This broad range of applications illustrates the importance of the ESP in industry. We note, however, that this list of other applications of the ESP is far from exhaustive. We defer discussion of other applications of the ESP (i.e., outside of the mining application) and similarly structured problems from the wider OR literature, where modelling and solution approaches may prove relevant, to Sect. 3.3.

In the literature on mining, equipment selection is a similar problem to mining method selection; and in the literature on construction, it is similar to shovel-truck productivity. The mining method selection problem is an approach to equipment selection that is based on the premise that the environmental conditions dictate the mining method used, and that the selection of truck and loader types follows intuitively from the mining method adopted. To simplify the task of selecting equipment

Table 3.1 This table categorises the problems solved in the related mining and construction literature (A–F)

Mining- and construction-related literature

	Background	Equipment selection	Mining method selection	Shovel-truck productivity	Scheduling	Dispatching and allocation
Alarie and Gamache [2]	×					
Amirkhanian and Baker [3]		×				
Başçcetin and Kesimal [12]		×				
Basçetin [14]		×				
Basçetin et al. [13]		×				
Bandopadhyay and Nelson [8]		×				
Bandopadhyay and Venkatasubramanian [9]		×				
Bazzazi et al.[16]		×				
Bitarafan and Ataei [22]			×			
Blackwell [23]	×			×		
Bozorgebrahimi et al. [25]	×					
Burt [27]		×				
Burt et al. [28]		×				
Burt et al. [30]		×				
Burt and Caccetta [31]	×					

(continued)

Table 3.1 (continued)

Mining- and construction-related literature	Background	Equipment selection	Mining method selection	Shovel-truck productivity	Scheduling	Dispatching and allocation
Caccetta and Hill [32]					×	
Celebi [36]		×				
Cebesoy [34]	×					
Cebesoy et al. [35]		×				
Czaplicki [43]	×					
Denby and Schofield [45]		×				
Douglas [47]				×		
Dunston et al. [48]	×					
Easa [49]						×
Edwards et al. [50]		×				
Eldin and Mayfield [51]		×				
Epstein et al. [52]					×	
Ercelebi and Kirmanli [54]	×					
Farid and Koning [55]		×				
Fricke [58]	×				×	
Frimpong et al. [59]					×	

Table 3.2 This table categorises the problems solved in the related mining and construction literature (G–Z)

Mining- and construction-related literature

	Background	Equipment selection	Mining method selection	Shovel-truck productivity	Scheduling	Dispatching and allocation
Ganguli and Bandopadhyay [62]		×				
Gleixner [64]					×	
Griffis, Jr [65]				×		
Halatchev [66]					×	
Hall and Daneshmend [67]	×					
Hassan et al. [69]		×				
Huang and Kumar [71]		×				
Ileri [72]		×				
Karelia Government [74]						×
Karshenas [76]				×		
Kesimal [77]				×		
Kumral and Dowd [80]					×	
Leontidis and Patmanidou [82]	×					
Markeset and Kumar [85]		×				
Marzouk and Moselhi [87]		×				
Marzouk and Moselhi [88]						×

(continued)

Table 3.2 (continued)

Mining- and construction-related literature	Background	Equipment selection	Mining method selection	Shovel-truck productivity	Scheduling	Dispatching and allocation
Michiotis et al. [89]		×				
Morgan [94]	×					
Morgan [95]	×					
Moselhi and Alshibani [96]		×				
Naoum and Haidar [97]		×				
Newman et al. [99]	×					
O'Hara and Suboleski [100]	×					
O'Shea [101]				×		
Schexnayder et al. [111]	×					
Smith et al. [112]				×		
Smith et al. [113]	×					
Ta et al. [115]						×
Tan and Ramani [116]					×	
Tomlingson [117]		×				
Topal and Ramazan [118]		×				
Wei et al. [121]			×			
Xinchun et al. [123]		×				
Zhongzhou and Qining [126]	×					

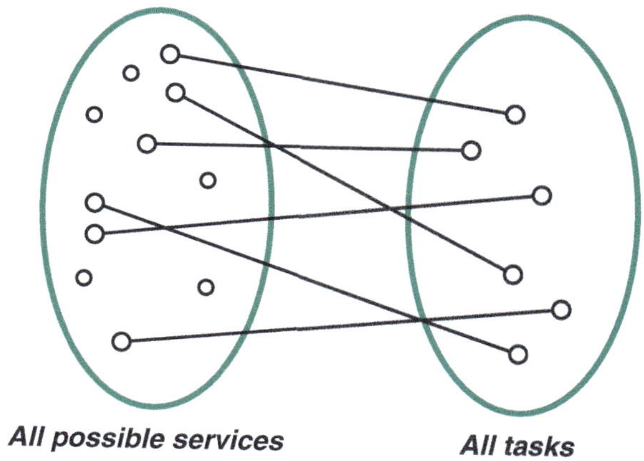

Fig. 3.1 The allocation and dispatching problems are concerned with matching services (e.g., provided by equipment) to tasks (e.g., moving ore)

while also selecting the mining method, solution approaches to this problem generally focus on choosing the correct excavation method for the given conditions. The shovel-truck productivity research area focuses on estimating and optimising the productivity of a truck and loader fleet. This literature generally relies on the notion that improving productivity translates into cost reductions [111]. However, the number of trucks performing the materials handling task affects the efficiency of the truck fleet [2]. Therefore, these methods extend in a simple way to find feasible solutions for the ESP.

Solution of the mining method selection problem is a preliminary step to solving the ESP, whereby mining engineers choose an appropriate excavation method based on environmental conditions. Early work on this problem (e.g. [23, 77, 116]), describes this approach in combination with a "match factor" (described in Sect. 3.3), as part of the procedure to select equipment.

Dispatching and allocation are also related topics in the literature on mining. The problem is to allocate tasks to equipment (see Fig. 3.1), which is a component of the ESP. In the scope of timed services, this becomes the dispatching problem. The key difference between dispatching and ESPs is that dispatching concerns generating a feasible online schedule for daily operational decisions, while the ESP concerns generating a purchase and salvage policy that is robust to the tactical planning level. The literature on allocation focuses on the satisfaction of productivity requirements, often with complex features such as bottleneck prevention; the literature on dispatch optimisation seeks to maximize the efficiency of the fleet at hand [99].

The ESP is related to asset management, where related subproblems in this category include equipment costing [82, 94, 100], production sequencing [66], facility equipment and machine selection in manufacturing systems [37], network planning [46], and equipment replacement [98, 108, 117].

Studying the similarity between various problems (or the structure they exhibit) is important to adequately solve a difficult problem such as the ESP. In the next section, we review both modelling and solution approaches for the ESP in mining, construc-

Table 3.3 This table categorises the problem structure in related OR literature

Problem structure	Network design	Vehicle routing problem	Equipment selection	Hub location	Scheduling scheduling	Assignment problem
Anderson et al. [4]	×					
Anderson et al. [5]	×					
Armacost et al. [6]	×					
Baldacci et al. [7]		×				
Barnhart et al. [11]	×					
Barnhart and Schneur [10]	×					
Baxter et al.[15]			×			
Bennett and Yano [17]			×			
Bienstock and Güunlük [19]	×					
Bienstock and Güunlük [20]	×					
Bienstock and Muratore [21]	×					
Boland et al. [24]				×		
Büdenbender et al. [26]	×					
Caramia and Guerriero [33]	×					
Chen [37]			×			
Cohn [39]	×					
Cordeau et al. [40]		×				
Crainic et al. [41]	×					
Croxton et al. [42]	×					
Dahl and Derigs [44]	×					
Derigs et al. [46]					×	

(continued)

Table 3.3 (continued)

Problem structure						
	Network design	Vehicle routing problem	Equipment selection	Hub location	Scheduling scheduling	Assignment problem
Equi et al. [53]					×	
Frangioni and Gendron [56]	×					
Galiano et al. [60]	×					
Gambardella et al. [61]					×	
Gendron et al. [63]	×				×	
Hane et al. [68]	×					×
Khan [78]			×			
Kim [79]	×				×	
Mirhosseyni and Web [90]			×			
Mitrović-Minić et al. [91]	×	×				
Moccia et al. [92]		×				
Montemanni and Gambardella [93]	×			×		
[98]			×			
Nassar [104]	×				×	
Pedersen and Crainic [105]	×					
Powell and Sheffi [107]	×					
Raack et al. [108]			×			
Rajagopalan [109]			×			
Raman et al. [110]		×				
Savelsbergh and Sol [114]	×					
Sung and Song [119]				×		
van Dam et al. [120]			×			
Webster and Reed [124]					×	

tion, and wider literature, and those approaches for similarly structured problems (or subproblems of the ESP).

3.3 Modelling and Solution Approaches

The problem structures that we address in this section include equipment selection, network design, vehicle routing, hub location, scheduling, and allocation. In Table 3.3, we provide a table of the problem structure in the related literature. However, we have organized the literature into modelling and solution approach subcategories to illustrate the success of some approaches (in some applications) and to highlight the emerging opportunity, in some cases, to apply these advancements to the ESP for surface mining. The order of the text is as follows: we begin by discussing the least sophisticated approaches, move towards exact approaches and search techniques that can complement or enhance an exact solution, and finish with solution verification approaches.

Markeset and Kumar [85] and Bozorgebrahimi et al. [25] each present life cycle costing (LCC) as an equipment selection method. LCC is a method for determining the cost per utilised hour (i.e., equipment utilised cost) of equipment if the equipment operates for its entire life span. A basic comparison can be made between each equipment utilised cost to determine the cheapest piece of equipment, although these comparisons do not tend to take into account the task to be performed or the time required to perform it. This type of analysis may be useful in determining a cost per hour for equipment, especially in a model that does not permit salvage (i.e., the equipment is kept for its entire life cycle). Some literature also uses cost estimation for truck transportation problems in which the focus on uncertain parameters aims to improve robustness of the solutions (e.g., [125]).

Heuristic or approximate methods and their use persist in industry. Heuristics can find feasible solutions quickly. However, some examples employ spreadsheets to aid manual iteration over a small subset of possibilities (see [51]). Another heuristic is an extension of the match factor ratio. The match factor ratio is an important productivity index in the mining industry (see Fig. 2.2). The match factor is simply the ratio of truck arrival rate to loader service time. Literature for the construction industry, in particular, uses match factor to determine a suitable truck fleet size. Smith et al. [113] recommend using the match factor formula as a means of determining the appropriate fleet size. However, an expert must select the best types of equipment before applying the formula. Smith et al. [113] reported that, at the time of publication, the earthmoving industry still used this ratio to determine an appropriate truck fleet size once the loader fleet and truck type have been established. Complete ESP solutions are typically obtained by applying match factor or mathematical programming approaches to determine the minimum number of trucks required for a mine plan (see [2]) and then using dynamic programming to determine allocation to mining locations (see [23]). Burt and Caccetta [31] extend the formula to account for heterogeneous fleets and multiple truck cycle times. We provide these extensions in Chap. 4.

Uncertainty in some parameters can lead to infeasibility of the truck allocation solution. Ta et al. [115] developed a stochastic model that incorporates real-time data for allocation of the fleet. Karimi et al. [75] addressed the uncertainty in parameters with a fuzzy optimisation allocation model, but their approach ignores the fixed charge (incurred at purchase), and thereby does not address the ESP as we define it here. In another example, Easa [49] developed two quadratic programming models for earthwork allocation. These models only allow for linear cost functions, as opposed to the more common piecewise linear cost functions. Chen [37] examines a multiperiod ESP model without transportation networks and develops a heuristic to address the difficulty arising from the multiperiod nature of the model. The authors use Lagrangian relaxation to provide bounds on the quality of their heuristic solutions.

A number of models incorporate net present value (NPV) analysis to allow comparisons between present and future cash flow. Typically, a multiplier (incorporating interest rates and depreciation) as a function of time can be appended to a cost-based objective function [28]. However, future interest rates are uncertain and difficult to predict. Wiesemann et al. [122] proposed a global optimisation model for accurate NPV under uncertainty, and solve it using a branch-and-bound-based heuristic.

Edwards et al. [50] used a linear programming model for selecting equipment in which the equipment is to be hired instead of purchased. However, the authors neglected to define the variables and explain how continuous variables could lead to integer values of equipment as a solution. That is, equipment is discrete in nature and a fraction of a piece of equipment cannot be hired. Land and Doig [81] established that simply rounding discrete variables from a linear program can lead to a violation of important discrete variable constraints or a solution in which the rounded variable values are vastly different from their optimal integer values.

Queuing theory was first notably applied to shovel-truck productivity by O'Shea [101]. In this work, O'Shea used queuing theory to predict the productivity of trucking fleets in an attempt to account for the productivity lost when the trucks queue at a loader. Much later, Karshenas [76] outlined several improvements and subsequently incorporated them into an equipment capacity selection model. This is a nonlinear optimisation model with a single constraint, and can be solved using direct search algorithms for global optimisation.

Griffis, Jr [65] developed a heuristic for determining the truck fleet size using queuing theory. This extended the work of O'Shea [101] for calculating the productivity of different fleet options by modelling the truck arrival rates as a Poisson process. Here, the authors assume that the time between arrivals follows an exponential distribution. Independence between arrivals is also a key assumption. Later, Farid and Koning [55] used simulation to verify the effectiveness of the equipment selection results of Griffis, Jr [65]. However, equipment bunching may violate the independence assumption. Bunching theory is the study of the bunching effect that can occur when equipment travels along the same route. Because trucking equipment does not travel at precisely the same speed (and therefore maintain uniformly distributed cycle times), equipment may cluster behind slower trucks, creating the bunching effect. Douglas [47], Morgan [95], and Smith et al. [112] describe equipment bunching in the context of a surface mine. However, the literature has thus far not included bunching in the modelling process. Instead, the aforementioned mining literature adopted shrinking factors to account for bunching, although bunching is a function of the quantity of equipment, the type of road, and many other factors.

Huang and Kumar [71] have extended this work in an attempt to select the size of the trucking fleet using a more accurate productivity estimate. They developed a fleet size selection queuing model to minimise the cost of idle machinery. Their model recommends selecting fleet sizes that match the maximum efficiency for both location and haulage equipment. Although using a productivity-focused objective function may not improve the economic result (e.g., by lowering the overall cost of materials handling), it is useful to consider the variability in some of the parameters of the ESP, such as truck cycle times and queue length. In production materials handling research, Raman et al. [109] used queuing theory to determine the optimal quantity of equipment in a transportation context, given a schedule.

Exact methods, such as integer programming, have provided an important methodology for equipment selection in surface mining. Network design models, in particular, capture the selection and flow aspects that are crucial to a good ESP model. In the mining literature, basic integer programming models are common. Simplifying assumptions reduces the problem instantiations to easily solvable cases. For example, nonlinear operating costs can be discretized to piecewise linear functions using age brackets, as in Burt [27], and Topal and Ramazan [118]. Cebesoy et al. [35] developed a systematic decision-making model for the selection of equipment types. They solved their model with a heuristic that uses a binary integer program in the final step. This model considers a single-period, single-location mine with homogeneous fleets. They perform compatibility matching of the equipment separately before solving the model.

In another example, Michiotis et al. [89] used a pure binary programming model for selecting the number, type, and locality of excavating equipment to work in a pit. The authors therefore ignore the transportation aspect of the problem. The model minimises the time to extract the material. In this model, the solution space is restricted by knapsack-based constraints that ensure that equipment is suitable for the size and shape of the bench, and for production requirements. Burt et al. [28] developed a mixed-integer programming model for equipment selection with a single source and destination. We provide these developments in Chap. 6. This model focused on the complex side constraints arising from heterogeneous fleets and the compatibility of the equipment. Outside of mining, Burt et al. [15] considered the ESP in the context of forestry harvesting, also using mixed-integer programming. This problem is similar to the surface mining problem, whereby the model selects the equipment and the number of hours of operation for a given harvesting schedule with respect to an underlying transportation problem. The authors have modeled the number of hours of operation so that the objective function is more accurate than the current standard in surface mining. That is, because the efficiency and cost of operating equipment changes with the age of the equipment (e.g., the number of hours the equipment has been used), including the age of the equipment in the objective function is practical.

Because an aspect of the ESP is a multicommodity network flow problem, it is useful to consider literature that focuses on this problem. Papers that provide a deep discussion of the structure, computational issues, solution approaches, and application of capacitated multicommodity network flow include Bienstock and Günlük [19, 20], Barnhart et al. [11] and Moccia et al. [92]. In an example of a combined network design and equipment selection problem, Anderson et al. [5] incorporated equipment selection into their intermodal transportation problem quite simply by adding variables to select equipment, and re-indexing the flow variables to account for the different types of possible equipment.

In the OR literature, Equi et al. [53] model the scheduling problem in the context of transportation using mixed-integer programming, and develop a Lagrangian relaxation solution approach. Other examples of Lagrangian relaxation in the context of network planning include Gendron et al. [63], Galiano et al. [60] and Zhang [125].

Because including a time index on a variable is important for the NPV costing, the quantity of variables in discrete models can sometimes become overwhelming. Reformulation is common in a bid to find a less naive and inhibitive way to capture the problem than, for example, the most obvious formulation. Good examples of network reformulations in this context include Armacost et al. [6], Cohn [39] and Frangioni and Gendron [56]. These papers each use composite binary variables to represent multiple decisions to simplify the model and to reduce its size. The papers then exploit the composite variable formulation in a decomposition approach; we describe the latter in the next paragraph. The composite variables capture overarching decisions, and a linear program can solve the underlying transportation problems. Kim [79] provides a discussion and comparison of some types of reformulations, such as node-arc versus path and tree formulations. Another possible approach is to use a set-partitioning model, such as in Baldacci et al.[7]. The authors consider the set of all feasible routes and partition them into sets that cover each customer's demand. They then construct heuristics to find good bounds on the optimal solution and use exact methods (such as branch and bound) to try to improve on the resulting optimality gap.

Decomposition approaches are widely used in the broader literature for problems that are too difficult to solve in complete form or for problems that are naturally composed of easy-to-solve subproblems. Dynamic programming, branch and bound, and Dantzig-Wolfe and Bender's decomposition are classic examples of decomposition approaches. Papers related to network planning that employ decomposition include Powell and Sheffi [105], Barnhart and Schneur [10], Mamer and McBride [57, 84]. Customizing the branching process is sensible for a problem with such an inherent structure as the ESP. Notably, the solution from one period depends on the solution from a previous period. In addition, the material flows imply the equipment solution. A typical approach in network planning applications is to develop a custom branch-and-cut algorithm, as in Croxton et al. [42], Baldacci et al. [7], and Cordeau et al. [40].

Bennett and Yano [17] describe a single-period equipment selection model with an underlying transportation problem. They adopt a Benders decomposition approach by observing the natural partitioning of the problem into equipment choice and service provision to satisfy the flow of product. Derigs et al. [46] address air cargo network planning, which involves flight selection, aircraft rotation, and cargo routing. This application is closely related to the service selection, service frequency, and equipment allocation aspects of equipment selection in surface mining. However, this problem involves additional complexities, such as crew scheduling and maintenance scheduling. The authors develop a column generation solution approach to combat the size of the problem (i.e., the number of variables required to express a practical instance). In a column generation approach, the columns represent feasible solutions in the problem. The key to this approach is to devise efficient heuristics for adding columns to the model. The overarching goal is to keep the search space minimal; therefore, this approach can be effective for problems that have an overwhelming number of variables or have an exploitable structure. Lübbecke and Desrosiers [83] provide a review of relevant techniques in column generation.

Fleet assignment or allocation has been widely considered in the mining literature, mostly because of the ease of the heuristic approach for (1) determining the equipment types, then (2) the fleet size, and subsequently (3) the fleet assignment. This problem is similar to the ESP when a mining schedule already exists (with the difference lying in the purchase and salvage requirements of the ESP). Webster and Reed [120] proposed a quadratic integer programming model that allocates material handling tasks to a single piece of equipment. This model allocates equipment rather than selecting the types and number of equipment, and is restricted to a single period. However, Hassan et al. [69] extended Webster and Reed [120]'s model to combine the equipment selection problem with the allocation problem. This model minimises the cost of operating the fleet subject to a knapsack and linking constraint set. In the broader literature, Hane et al. [68] provides another example of fleet assignment in the context of complex networks. They model a fleet assignment problem as a multicommodity flow problem with side constraints. Some considerations in their paper, such as defining the problem on a time-expanded network, are particularly relevant for the ESP in the context of mining. They developed a specialized branch-and-cut algorithm based on the structure of the problem.

Preprocessing techniques are an important part of solving mixed-integer programs, particularly in the presence of symmetry (arising, for example, from representing identical equipment with separate variables) and excessive quantities of binary variables in the discrete description of the problem. These techniques are not common in the mining literature, although Burt [27] provides a brief description of variable and constraint reduction. Other preprocessing examples in related literature include Ileri [72] who preprocesses by observing dominance among route assignments, and Boland et al. [24] who provide multiple properties for preprocessing flow variables and constraint reduction.

The OR literature includes an increasing number of cases of local search techniques used to improve the efficiency of exact algorithms, and as stand-alone heuristics. Zhang [125] considered a less-than-load planning problem with the assumption that freight flow patterns repeat, thus considerably reducing the number of commodity variables. However, the author also argued for smaller time steps than is common in the literature to reduce the variance in travel times. Other local search techniques in the context of the capacitated network planning problem are proposed in Büdenbender et al. [26], Sung and Song [114], Hewitt [70], and Caramia and Guerriero [33].

Artificial intelligence techniques are prevalent in large-scale mining applications because of their ability to find feasible solutions within a comparatively short time Clèment and Vagenas [38]. The most common methods in the literature are the decision support system methods [9, 45], and genetic algorithms [86, 87, 96, 97, 123]. Various decision support tools, such as the analytical hierarchy process [14] and expert systems [3], apply priorities to decisions for logic-based heuristic solutions. These methods consider the entire process of equipment selection holistically, including site conditions, geology, environment, and equipment matching. Equipment matching is a step beyond merely considering compatibility, where ranks (formed in a preprocessing step) represent the suitability of pairs.

Genetic algorithms are a heuristic solution technique that evolve a solution after several generations of stochastic selection based on a fitness criterion. Numerous examples apply genetic algorithms to the ESP. Naoum and Haidar [97] developed a genetic algorithm model for the problem. In their model, they incorporate the lifetime-discounted cost of the equipment, which arises from the assumption that the equipment operates from purchase until its official retirement age and is not sold or replaced before that time. Moselhi and Alshibani [96] developed a genetic algorithm to choose equipment for a single-location, single-period mining schedule.

The complex interplay between types of equipment has led to literature that focuses on attribute matching, such as Abdel-Malek and Resare [1] in the production research literature and Bazzazi et al. [16] in the mining literature. Attribute-based selection methods include multiattribute decision-making modelling [8, 13] and fuzzy set theory [12, 22, 121]. Fuzzy programming approaches may help to combat the uncertain nature of some of the data.

The basic attribute-matching problem can select the equipment over multiple periods. Ganguli and Bandopadhyay [62] also developed an expert system for equipment selection. However, their method requires the user to input the relative importance of the factors, which is typically difficult to quantify and substantiate. Khan [78] developed a knowledge-based heuristic to focus on attribute matching. Mirhosseyni and Web [90] presented a combined expert system and genetic algorithm approach for the selection and assignment of equipment for materials handling (not only for surface mining applications).

Finally, simulation approaches can verify solutions (obtained from other methods) for robustness and quality, and they can also be used to obtain solutions. For example, Marzouk and Moselhi [88] designed a model using simulation and genetic algorithms to trade off two objectives, time and cost, for the construction industry.

We provide a summary of these modelling and solution approaches in Tables 3.4 and 3.5.

Table 3.4 This table categorises the OR methods used in each paper, A–G

Methods	Mixed-integer and linear programming	Artificial intelligence	Heuristic	Simulation	Queueing
Anderson et al. [4]	×				
Anderson et al. [5]	×				
Armacost et al. [6]	×				
Bascetin et al. [13]		×			
Baldacci et al. [7]	×				
Bandopadhyay and Nelson [8]		×			
Bandopadhyay and Venkatasubramanian [9]		×			
Barnhart et al. [11]	×				
Barnhart and Schneur [10]	×				
Baxter et al. [15]	×				
Bazzazi et al. [16]		×			
Bennett and Yano [17]	×				
Bienstock [18]	×				
Bienstock and Günlük [19]	×				
Bienstock and Günlük [20]	×				
Bienstock and Muratore [21]	×				
Bitarafan and Ataei [22]		×			
Boland et al. [24]	×				
Büdenbender et al. [26]	×	×			
Burt [27]	×				
Burt et al. [28]	×				

(continued)

Table 3.4 (continued)

Methods					
	Mixed-integer and linear programming	Artificial intelligence	Heuristic	Simulation	Queueing
Caccetta and Hill [32]	×				
Caramia and Guerriero [33]			×		
Cebesoy et al. [35]		×			
Chen [37]			×		
Cohn [39]	×				
Cordeau et al. [40]	×				
Crainic et al. [41]		×			
Croxton et al. [42]	×				
Dahl and Derigs [44]		×		×	
Denby and Schofield [45]		×			
Derigs et al. [46]		×			
Easa [49]			×		
Edwards et al. [50]	×				
Eldin and Mayfield [51]			×		
Equi et al. [53]	×				
Farid and Koning [55]				×	
Frangioni and Gendron [56]	×				
Frangioni and Gendron [57]	×				
Fricke [58]	×				
Frimpong et al. [59]		×		×	
Galiano et al. [60]	×				
Gambardella et al. [61]		×		×	

Table 3.5 This table categorises the OR methods used in each paper, G–Z

Methods	Mixed-integer and linear programming	Artificial intelligence	Heuristic	Simulation	Queueing
Ganguli and Bandopadhyay [62]		×			
Gendron et al. [63]	×				
Gleixner [64]	×				
Griffis, Jr [65]		×			
Hane et al. [68]	×				
Hewitt [70]			×		
Huang and Kumar [71]					×
Ileri [72]			×		
Irnich [73]	×				
Karelia Government [74]		×			
Karshenas [76]					×
Khan [78]			×		
Kim [79]	×				
Kumral and Dowd [80]		×			
Land and Doig [81]	×				
Lübbecke and Desrosiers [83]	×				
Mamer and McBride [84]	×				
Marzouk and Moselhi [86]		×			
Marzouk and Moselhi [88]		×		×	
Michiotis et al. [89]	×				
Mirhosseyni and Web [90]		×			
Mitrović-Minić et al. [91]			×		

(continued)

Table 3.5 (continued)

Methods	Mixed-integer & linear programming	Artificial intelligence	Heuristic	Simulation	Queueing
Moccia et al. [92]	×				
Montemanni and Gambardella [93]	×				
Moselhi and Alshibani [96]		×			
Naoum and Haidar [97]		×			
Nassar [98]				×	
O'Shea [101]					×
Pedersen and Crainic [104]	×				
Powell and Sheffi [105]	×				
Raack et al. [107]	×				
Raman et al. [109]					×
Sung and Song [114]			×		
Tan and Ramani [116]	×				
Topal and Ramazan [118]	×				
van Dam et al. [119]		×			
Webster and Reed [120]	×				
Wei et al. [121]		×			
Xinchun et al. [123]		×			
Zhang et al. [124]			×		
Zhongzhou and Qining [126]					×

3.4 Conclusion

Heuristic methods, including life-cycle costing, are the simplest to implement of all the approaches. The solution process is typically also easy to understand. From this standpoint, these approaches are practical for mining engineers. Practicability is a desirable attribute, because the quantity of parameters necessary to construct the problem and the different time scales in decisions makes the problem seem overwhelmingly complicated. The queuing theory, artificial intelligence, and simulation literature try to address these complexities in an efficient and easy-to-understand way, but lack the solving power to deal with the number of decisions that must be made across different time horizons. In the literature, the preference for exact approaches, and in particular, mixed-integer programming (MIP) is clear. This may be because MIP is capable of handling larger-scale models of the problem, such as multiple scheduling and purchasing periods, heterogeneity of fleets, and other complex side constraints. The treatment of the ESP in the context of these methods has, however, been weak. The mining literature addresses only overly simplified instantiations of the problem and fails to sufficiently address the need for robustness in the solution. However, the OR literature has many modelling and solution tools that may improve the solution of the ESP. Many difficult, similarly structured, and large-scale problems have been solved using exact methods. Furthermore, computationally difficult methods, such as MIP approaches, can provide measurably good-quality solutions efficiently.

Clearly, the literature includes very broad definitions of the truck and loader equipment selection problem for surface mining. In some research, the truck and loader types are selected deterministically and the fleet sizes implied from production requirements. In more recent work, the complex interplay between different types of equipment is addressed. The gap between these two solution formats indicates an evolution in what could be considered a suitable definition of the problem. However, additional progress can still be made on this problem, because the uncertainty in the input parameters have not yet been properly addressed. Although some uncertainty may only affect the optimality of a solution, such as depreciation, interest rates, and fuel prices, other inputs affect the feasibility of the fleet, such as truck cycle time, availability, bunching, and truckload variability. The uncertainty of the latter set of parameters must be addressed first in new research. To achieve this using the exact algorithmic approach, the problem and subsequent model must be controlled by complementary techniques, such as those we describe in the following paragraphs.

Preprocessing is of clear benefit because of the dependency of solutions on previous periods, but one must avoid destroying structure that could be exploited in decomposition techniques. Approximation algorithms and heuristics can obtain good initial solutions that can then initialize a branch-and-bound algorithm to improve computation time. One approach could be to focus on solving the underlying transportation problem (with approximation heuristics) and then infer the required selection. To this end, one could use the approximation algorithm in Bienstock [18]. Alternative heuristics include Savelsbergh and Sol [110] and Zhang et al. [124]. Tabu-search and agent-based methods may also provide good starting solutions, as in Crainic et al. [41] and van Dam et al. [119].

Separation procedures available in the literature could be computationally advantageous in a branch-and-cut approach, such as those in Bienstock and Muratore [21], Irnich [73] and Raack et al. [107]. Constructing minimal cover cuts from the productivity constraint (in knapsack constraint form, such as in [24]); cut-sets on the flow (as in Boland et al. [21]); reachability sets on the flow (as in [93]); and lifting on precedence constraints (as in [40]), could lead to computational improvement. Commercial solvers implement some of these for the general case.

Because of the issues arising from the time fidelity, a rolling horizon could be practical. Mitrović-Minić et al. [91] provide such an example for network planning. The time fidelity discrepancy may reduce if the schedule is cyclic, as in [4, 125], although this might lead to an important loss of detail in the model.

The most important focus for future research is to generate robust solutions. This could mean considering uncertain parameters in the modelling process, or generating solutions that are robust against unlikely events. Of particular importance is the need to account for uncertainty in the key parameters. Starting points include Ileri [72]. One other potential aspect is incorporating bunching into future modelling approaches. Other approaches include scenario generation, as in Pedersen and Crainic [104] and Preuss and Hellingrath [106].

In conclusion, open-pit mining involves the removal of a large quantity of material, requiring extraction and hauling equipment. The 50 years of publications on this problem demonstrate a colored and adventurous collection of modelling and solution approaches. In spite of this, satisfactory robust solutions are missing from the literature. To direct future applied research, we have highlighted OR literature illustrating advancements that may aid in an efficient solution of the ESP. Ultimately however, the goal is to embed the equipment selection problem into the long-term mine plan such that they can be optimised holistically.

Acknowledgements Components of this chapter have been published in [29]

References

1. L. Abdel-Malek, L.J. Resare, Algorithm based decision support system for the concerted selection of equipment in machining/assembly cells. Int. J. Prod. Res. **38**(2), 323–339 (2000)
2. S. Alarie, M. Gamache, Overview of solution strategies used in truck dispatching systems for open pit mines. Int. J. Surf. Min. Reclam. Environ. **16**(1), 59–76 (2002)
3. S. Amirkhanian, N. Baker, Expert system for equipment selection for earth-moving operations. J. Construction Eng. Manag. **118**(2), 318–331 (1992)
4. J. Anderson, T.G. Crainic, M. Christiansen, Service network design with asset management: formulations and comparative analyses. Transportation Res. Part C Emerg. Technol. **17**(2), 197–207 (2009a)

5. J. Anderson, T.G. Crainic, M. Christiansen, Service network design with management and coordination of multiple fleets. Eur. J. Oper. Res. **193**, 377–389 (2009b)
6. A. Armacost, C. Barnhart, K. Ware, Composite variable formulations for express shipment service network design. Transportation Sci. **36**(1), 1–20 (2002)
7. R. Baldacci, N. Christofides, A. Mingozzi, An exact algorithm for the vehicle routing problem based on the set partitioning formulation with additional cuts. Math. Program. **115**(2), 351–385 (2008)
8. S. Bandopadhyay, M. Nelson, Multiple-criteria group decision support systems with application to a surface mine equipment selection problem, in *Computers in the Mineral Industry*, ed. by R.V. Ramani, B.K. Mozumdar, A.B. Samaddar (Balkema, Rotterdam, 1994), pp. 111–126
9. S. Bandopadhyay, P. Venkatasubramanian, Expert system as decision aid in surface mine equipment selection. Int. J. Surf. Mining **1**, 159–165 (1987)
10. C. Barnhart, R.R. Schneur, Air network design for express shipment service. Oper. Res. **44**(6), 852–863 (1996)
11. C. Barnhart, N. Krishnan, D. Kim, K. Ware, Network design for express shipment delivery. Comput. Optim. Appl. **21**(3), 239–262 (2002)
12. A. Başçetin, A. Kesimal, The study of a fuzzy set theory for the selection of an optimum coal transportation system from pit to the power plant. Int. J. Surf. Min. Reclam. Environ. **13**(3), 97–101 (1999)
13. A. Başçetin, O. Oztaş, and A. I. Kanlı, A new developed software for equipment selection in mining engineering, in *Mine Planning and Equipment Selection 2004 : Proceedings of the Thirteenth International Symposium on Mine Planning and Equipment Selection*, ed. by M. Hardygora, G. Paszkowska, M. Sikora, Wroclaw, Poland, 1–3 September, London (2004), pp. 527–536. A.A. Balkema. ISBN 0415359376
14. A. Başçetin, An application of the analytic hierarchy process in equipment selection at Orhaneli open pit coal mine. Mining Technol. Trans. Intitute Mining Metallurgy Part A **113**, A192–A199 (2004)
15. M. Baxter, M. Brown, H.S. Gan, A decision support tool for equipment replacement in forestry harvesting operations, in *Proceedings of the 45th Annual Conference of the ORSNA*, pp. 363–372, https://secure.orsnz.org.nz/conf45/index.php. Accessed 24 December 2010
16. A.A. Bazzazi, M. Osanloo, B. Karimi, Optimal open pit mining equipment selection using fuzzy multiple attribute decision making approach. Archives Min. Sci. **54**(2), 301–320 (2009)
17. D.P. Bennett, C.A. Yano, A decomposition approach for an equipment selection and multiple product routing problem incorporating environmental factors. Eur. J. Oper. Res. **156**(3), 643–664 (2004)
18. D. Bienstock, *Potential Function Methods for Approximately Solving Linear Programming Problems: Theory and Practice* (Kluwer Academic, 2001)
19. D. Bienstock, O. Günlük, Computational experience with a difficult mixed-integer multicommodity flow problem. Math. Program. **68**(1–3), 213–237 (1995)
20. D. Bienstock, O. Günlük, Capacitated network design-polyhedral structure and computation. Informs J. Comput. **8**, 243–259 (1996)
21. D. Bienstock, G. Muratore, Strong inequalities for capacitated survivable network design problems. Math. Program. **89**, 127–147 (2000)
22. M. Bitarafan, M. Ataei, Mining method selection by multiple criteria decision making tools. J. South African Inst. Min. Metallurgy **104**(9), 493–498 (2004)
23. G.H. Blackwell, Estimation of large open pit haulage truck requirements. CIM Bull. **92**(1028), 143–148 (1999)
24. N. Boland, M. Krishnamoorthy, A. Ernst, J. Ebery, Preprocessing and cutting for multiple allocation hub location problems. Eur. J. Oper. Res. **155**(3), 638–653 (2004)
25. A. Bozorgebrahimi, R. Hall, M. Morin, Equipment size effects on open pit mining performance. Int. J. Surf. Min. Reclam. Environ. **19**(1), 41–56 (2005)
26. K. Büdenbender, T. Grünert, H.-J. Sebastian, A hybrid tabu search/branch-and-bound algorithm for the direct flight network design problem. Transportation Sci. **34**(4), 364–380 (2000)

27. C. Burt, *An Optimisation Approach to Materials Handling in Surface Mines*. Ph.D. thesis, Mathematics and Statistics, Curtin University of Technology, Bentley, Perth, Australia (2008). http://espace.library.curtin.edu.au:80/R?func=dbin-jump-full&local_base=gen01-era02&object_id=165737
28. C. Burt, L. Caccetta, P. Welgama, L. Fouché, Equipment selection with heterogeneous fleets for multiple period schedules. J. Oper. Res. Soc. **62**, 1498–1509 (2011)
29. C.N. Burt, L. Caccetta, Equipment selection for surface mining: a review. Interfaces **44**(2), 143–162 (2014)
30. C.N. Burt, L. Caccetta, L. Fouché, P. Welgama, An MILP approach to multi-location, multi-period equipment selection for surface mining with case studies. J. Ind. Manag. Optim. **12**(2), 403–430 (2016)
31. Christina Naomi Burt and Louis Caccetta, Match factor for heterogeneous truck and loader fleets. Int. J. Min. Reclam. Environ. **21**, 262–270 (2007)
32. L. Caccetta, S. Hill, An application of branch and cut to open pit mine scheduling. J. Global Optim. **27**, 349–365 (2003)
33. M. Caramia, F. Guerriero, A heuristic approach to long-haul freight transportation with multiple objective functions. Omega **37**(3), 600–614 (2009)
34. T. Cebesoy, Surface mining equipment cost analysis with a developed linear break even model. Int. J. Surf. Min. Reclam. Environ. **11**, 53–58 (1997)
35. T. Cebesoy, M. Gözen, S. Yahşi, A systematic decision making model for optimum surface mining equipment selection, in *Mine Planning and Equipment Selection 1995: Proceedings of the Fourth International Symposium on Mine Planning and Equipment Selection*, eds. by R.K. Singhal, A.K. Mehrotra, J. Hadjigeorgiou, R. Poulin, Calgary, Canada, 31st October-3 November 1995, pp. 369–377, Rotterdam, 1995. A.A. Balkema
36. N. Çelebi, An equipment selection and cost analysis system for openpit coal mines. Int. J. Surf. Min. Reclam. Environ. **12**, 181–187 (1998)
37. M. Chen, A heuristic for solving manufacturing process and equipment selection problems. Int. J. Prod. Res. **37**(2), 359–374 (1999)
38. S. Clèment, N. Vagenas, Use of genetic algorithms in a mining problem. Int. J. Surf. Min. Reclam. Environ. **8**(4), 131–136 (1994)
39. A.E. Cohn, *Composite-Variable Modeling for Large-Scale Problems in Transportation and Logistics*. Ph.D. thesis, Massachusetts Institute of Technology (2002)
40. J.-F. Cordeau, M. Iori, G. Laporte, J.J.S. Gonzalez, *A branch-and-cut algorithm for the pickup and delivery traveling salesman problem with LIFO loading* (Technical report, CIRRELT, 2007)
41. T.G. Crainic, M. Gendreau, J.M. Farvolden, A simplex-based tabu search method for capacitated network design. Informs J. Comput. **12**(3), 223–236 (2000)
42. K.L. Croxton, B. Gendron, T.L. Magnanti, Models and methods for merge-in-transit operations. Transportation Sci. **37**(1), 1–22 (2003)
43. J.M. Czaplicki, On a number of spare haulage units for the shovel-truck system in open pit mines. Int. J. Surf. Min. Reclam. Environ. **6**, 57–59 (1992)
44. S. Dahl, U. Derigs. Planning in express carrier networks: a simulation study. in *Operations Research Proceedings 2008: Selected Papers of the Annual International Conference of the German Operations Research Society (GOR)*, eds. by B. Fleischmann, K.-H. Borgwardt, R. Klein, University of Augsburg, September 3–5 2008, pp. 259–264 (Springer, Heidelberg, 2009)
45. B. Denby, D. Schofield, Applications of expert systems in equipment selection for surface mine design. Int. J. Surf. Min. Reclam. **4**(4), 165–171 (1990)
46. U. Derigs, S. Friederichs, S. Schäfer, A new approach for air cargo network planning. Transportation Sci. **43**(3), 370–380 (2009)
47. J. Douglas, Prediction of shovel-truck production: A reconciliation of computer and conventional estimates. Technical report no. 37, Stanford University, California (1964)
48. P.S. Dunston, J.V. Sinfield, T.Y. Lee, Technology development decision economics for real-time rolling resistance monitoring of haul roads. J. Construction Eng. Manag. **133**(5), 393–402 (2007)

49. S.M. Easa, Earthwork allocations with linear unit costs. J. Construction Eng. Manag. **114**(4), 641–655 (1988)
50. D. Edwards, H. Malekzadeh, S. Yisa, A linear programming decision tool for selecting the optimum excavator. Structural Surv. **19**(2), 113–120 (2001)
51. N. Eldin, J. Mayfield, Determination of most economical scrapers fleet. J. Construction Eng. Manag. **131**(10), 1109–1114 (2005)
52. R. Epstein, S. Gaete, F. Caro, A. Weintraub, P. Santibanez, J. Catalan, Optimizing long term planning for underground copper mines, in *Proceedings of the Copper 2003-Cobre 2003, 5th International Conference*, vol. I, Santiago, Chile, pp. 265–279. CIM and the Chilean Institute of Mining (2003)
53. L. Equi, G. Gallo, S. Marziale, A. Weintraub, A combined transportation and scheduling problem. Eur. J. Oper. Res. **97**(1), 94–104 (1997)
54. S. Erçelebi, C. Kirmanli, Review of surface mining equipment selection techniques, in [102], pp. 547–553
55. F. Farid, T.L. Koning, Simulation verifies queuing program for selecting loader-truck fleets. J. Construction Eng. Manag. **120**(2), 386–404 (1994)
56. A. Frangioni, B. Gendron, 0–1 reformulations of the multicommodity capacitated network design problem. Discrete Appl. Math. **157**(6), 1229–1241 (2009)
57. A. Frangioni, B. Gendron, *A stabilized structured Dantzig-Wolfe decomposition method* (Technical report, CIRRELT, 2010)
58. C. Fricke, *Applications of Integer Programming in Open Pit Mining*. Ph.D. thesis, Department of Mathematics and Statistics, University of Melbourne (2006)
59. S. Frimpong, E. Asa, J. Szymanski, Intelligent modeling: advances in open pit mine design and optimization research. Int. J. Surf. Min. Reclam. Environ. **16**(2), 134–143 (2002)
60. G. Galiano, T. Kaihara, G. Liotta, G. Stecca, A comprehensive model for short-haul and long-haul transportation, in *2010 8th IEEE International Conference on Industrial Informatics (INDIN)* (IEEE, 2010), pp. 939–944
61. L. Gambardella, M. Mastrolilli, A. Rizzoli, M. Zaffalon, Managing resource allocation, scheduling and simulation for an intermodal container terminal. J. Belgian Oper. Res. Soc. **40**(3–4), 195–209 (2002)
62. R. Ganguli, S. Bandopadhyay, Expert system for equipment selection. Int. J. Surf. Min. Reclam. Environ. **16**(3), 163–170 (2002)
63. B. Gendron, T. Crainic, A. Frangioni, Multicommodity capacitated network design, in *Telecommunications Network Planning*, ed. by B. Sansó, P. Soriano (Kluwer Academic Publishers, Dordrecht, 1998), pp. 1–19
64. A. Gleixner, Solving large-scale open pit mining production scheduling problems by integer programming. Master's thesis (Technische Universität, Berlin, 2008)
65. F.H. Griffis Jr., Optimizing haul fleet size using queueing theory. J. Construction Eng. Div. **94**(1), 41–54 (1968)
66. R. Halatchev, The time aspect of the optimum long-term open pit production sequencing, in *Application of computers and operations research in the mineral industry*, ed. by S. Bandopadhyay, pp. 133–146, Littleton, Colorado (2002). Society of Mining Engineers
67. R.A. Hall, L.K. Daneshmend, Reliability modelling of surface mining equipment: data gathering and analysis methodologies. Int. J. Surf. Min. Reclam. Environ. **17**(3), 139–155 (2003)
68. C. Hane, C. Barnhart, E. Johnson, R. Marsten, G. Nemhauser, G. Sigismondi, The fleet assignment problem: solving a large-scale integer program. Math. Program. **70**(1–3), 211–232 (1995)
69. M. Hassan, G. Hogg, D. Smith, A construction algorithm for the selection and assignment of materials handling equipment. Int. J. Prod. Res. **23**(2), 381–392 (1985)
70. M. Hewitt, *Integer programming based search*. Ph.D. thesis, Industrial and Systems Engineering, Georgia Institute of Technology (2009)
71. Y. Huang, U. Kumar, Optimising the number of load-haul-dump machines in a swedish mine, in *Mine Planning and Equipment Selection 1994: Proceedings of the third International Symposium on Mine Planning and Equipment Selection*, ed. by A.G. Pasamehmetoglu, Istanbul, Turkey, 18–20 October 1994, pp. 353–358, Rotterdam (1994). A.A. Balkema

72. Y. Ileri, *Drayage Optimization in Truck/Rail Networks*. Ph.D. thesis, Industrial and Systems Engineering, Georgia Institute of Technology (2006)
73. S. Irnich, *Netzwerk-Design für zweistufige Transportsysteme und ein Branch-and-Price-Verfahren für das gemischte Direkt- und Hubflugproblem*. Ph.D. thesis, Lehrstuhl für Unternehmensforschung, Rheinisch-Westfälische Technische Hochschule (2002)
74. Karelia Government. Cat 992g (2007), http://www.gov.karelia.ru/gov/News/2001/1002.jpg
75. S. Jamshid Mousavi, A. Kaveh, A. Afshar, Fuzzy optimization model for earthwork allocations with imprecise parameters. J. Construction Eng. Manag. **133**(2), 181–190 (2007)
76. S. Karshenas, Truck capacity selection for earthmoving. J. Construction Eng. Manag. **115**(2), 212–227 (1989)
77. A. Kesimal, Shovel-truck productivity and efficiency studies for overburden removal in an open-pit coal mine. Trans. Inst. Min. Metallurgy Sect. A Min. Industry **107**, A37–A40 (1998)
78. R. Khan, *Selection of material handling equipment*. Master's thesis, NWFP University of Engineering and Technology, Peshawar, Pakistan (2006)
79. D. Kim, *Large scale transportation service network design: models, algorithms and applications*. Ph.D. thesis, Department of Civil and Environmental Engineering, Massacheusetts Institute of Technology (1997)
80. M. Kumral, P. Dowd, A simulated annealing approach to mine production scheduling. J. Oper. Res. Soc. **56**, 922–930 (2005)
81. A. Land, A. Doig, An automatic method of solving discrete programming problems. Econometrica **28**(3), 497–520 (1960)
82. M. Leontidis, B. Patmanidou, Cost estimation of earthmoving done by hydraulic shovels and trucks in mines, in *Mine Planning and Equipment selection 2000: Proceedings of the Ninth International Symposium on Mine Planning and Equipment Selection /Athens/Greece/6-9 November 2000*, eds. by G.N. Panagiotou, T.N. Michalakopoulos, pp. 631–634, Rotterdam (2000). A.A. Balkema
83. M. Lübbecke, J. Desrosiers, Selected topics in column generation. Oper. Res. **53**(6), 1007–1023 (2005)
84. J. Mamer, R. McBride, A decomposition-based pricing procedure for large-scale linear programs: an application to the linear multicommodity flow problem. Manag. Sci. **46**(5), 693–709 (2000)
85. T. Markeset, U. Kumar, Application of LCC techniques in selection of mining equipment and technology, in [102], pp. 635–640
86. M. Marzouk, O. Moselhi, Simulation optimization for earthmoving operations using genetic algorithms. Construction Manag. Econ. **20**(6), 535–543 (2002a)
87. M. Marzouk, O. Moselhi, Selecting earthmoving equipment fleets using genetic algorithms, in *Proceedings of the 2002 Winter Simulation Conference* eds. by E. Yucesan, C. Chen, J. Snowdon, J. Charnes, vol. 2, pp. 1789–1796. IEEE Digital(2002b)
88. M. Marzouk, O. Moselhi, Multiobjective optimization of earthmoving operations. J. Construction Eng. Manag. **130**(1), 105–114 (2004)
89. A. Michiotis, D. Xerocostas, N. Galitis, A new integrated system for selecting mining equipment. Comput. Industrial Eng. **34**(2), 391–397 (1998)
90. S. Hamid, L. Mirhosseyni, P. Web, A hybrid fuzzy knowledge-based expert system and genetic algorithm for efficient selection and assignment of material handling equipment. Expert Syst. Appl. **36**, 11875–11887 (2009)
91. S. Mitrović-Minić, R. Krishnamurti, G. Laporte, Double-horizon based heuristics for the dynamic pickup and delivery problem with time windows. Transp. Res. **38**(8), 669–685 (2004)
92. L. Moccia, J.-F. Cordeau, G. Laporte, S. Ropke, M.P. Valentini, Modeling and solving a multimodal transportation problem with flexible-time and scheduled services. Networks **57**(1), 53–68 (2011)
93. R. Montemanni, L. Gambardella, Exact algorithms for the minimum power symmetric connectivity problem in wireless networks. Comput. Oper. Res. **32**(11), 2891–2904 (2005)
94. B. Morgan, Cost effective equipment applications zones, in [103], pp. 323–328
95. B. Morgan, Optimizing truck-loader matching, in [103], pp. 313–320

96. O. Moselhi, A. Alshibani, Optimization of earthmoving operations in heavy civil engineering projects. J. Construction Eng. Manag. **135**(10), 948–954 (2009)
97. S. Naoum, A. Haidar, A hybrid knowledge base system and genetic algorithms for equipment selection. Eng. Construction Architectural Manag. **7**(1), 3–14 (2000)
98. K. Nassar, Managing construction equipment buy and sell decisions replacement: a simulation game, in *Proceedings of the 37th Annual Conference on ASC Associated Schools of Construction*, pp. 187–198 (2001)
99. A. Newman, E. Rubio, R. Caro, A. Weintraub, K. Eurek, A review of operations research in mine planning. Interfaces **40**(3), 222–245 (2010)
100. T. O'Hara, S. Suboleski, Costs and cost estimation, in *SME Mining Engineering Handbook*, vol. 2, ed. by H.L. Hartman (Colorado, SME, Littleton, 1992), pp. 405–424
101. J.B O'Shea, An application of the theory of queues to the forecasting of shovel-truck fleet productions. Master's thesis, University of Illinois, Urbana, Illinois (1964)
102. G.N. Panagiotou, T.N. Michalakopoulos, eds. *Mine Planning and Equipment Selection 2000: Proceedings of the Ninth International Symposium on Mine Planning and Equipment Selection/Athens/Greece/6-9 November 2000*, Rotterdam (2000). A.A. Balkema
103. A.G. Pasamehmetoglu, ed. *Mine Planning and Equipment Selection 1994 : Proceedings of the Third International Symposium on Mine Planning and Equipment Selection, Istanbul, Turkey, 18-20 October 1994*, Rotterdam (1994). A.A. Balkema
104. M.B. Pedersen, T.G. Crainic, *Optimization of intermodal freight train service schedules on train canals* (Technical report, CIRRELT, 2007)
105. W.B. Powell, Y. Sheffi, Design and implementation of an interactive optimization system for network design in the motor carrier industry. Oper. Res. **37**(1), 12–29 (1989)
106. M. Preuss, B. Hellingrath. Tactical planning of sustainable transportation by logistics service providers for the automotive industry, in *Proceedings of the 43rd CIRP International Conference on Manufacturing Systems, Vienna, 26-28 May 2010*, pp. 252–262 (2010)
107. C. Raack, A. Koster, S. Orlowski, R. Wessäly, On cut-based inequalities for capacitated network design polyhedra. Networks **57**(2), 141–156 (2011)
108. S. Rajagopalan, Capacity expansion and equipment replacement: a unified approach. Oper. Res. **46**(6), 846–857 (1998)
109. D. Raman, S. Nagalingam, B. Gurd, G. Lin, Quantity of material handling equipment–a queuing theory based approach. Robotics Comput. Integr. Manufacturing **25**, 348–357 (2009)
110. M. Savelsbergh, M. Sol, Drive: dynamic routing of independent vehicles. Oper. Res. **46**(4), 474–490 (1998)
111. C. Schexnayder, S. Weber, B. Brooks, Effect of truck payload weight on production. J. Construction Eng. Manag. **125**(1), 1–7 (1999)
112. S.D. Smith, J.R. Osborne, M.C. Forde, Productivity estimation in back-acter/dump-truck earth-moving operations. Proc. Inst. Civil Eng. Transport **111**, 125–131 (1995)
113. S.D. Smith, G.S. Wood, M. Gould, A new earthworks estimating methodology. Construction Manag. Econ. **18**(2), 219–228 (2000)
114. C.S. Sung, S.H. Song, Integrated service network design for a cross-docking supply chain network. J. Oper. Res. Soc. **54**(12), 1283–1295 (2003)
115. C. Ta, J. Kresta, J. Forbes, H. Marquez, A stochastic optimization approach to mine truck allocation. Int. J. Surf. Mining Reclam. Environ. **19**(3), 162–175 (2005)
116. S. Tan, R.V. Ramani, Optimization models for scheduling ore and waste production in open pit mines, in *23rd Application of Computers and Operations Research in the Mineral Industry* ed. by Y.C. Kim, pp. 781–791, Littleton, Colorado (1992). Society for Mining, Metallurgy, and Exploration
117. P. Tomlingson, Equipment replacement considerations, in *Mine Planning and Equipment Selection 2000: Proceedings of the Ninth International Symposium on Mine Planning and Equipment Selection/Athens/Greece/6-9 November 2000* eds. by G.N. Panagiotou, T.N. Michalakopoulos, pp. 691–693, Rotterdam (2000). A.A. Balkema
118. E. Topal, S. Ramazan, A new MIP model for mine equipment scheduling by minimizing maintenance cost. Eur. J. Oper. Res. **207**(2), 1065–1071 (2010)

119. K.H. van Dam, Z. Lukszo, L. Ferreira, A. Sirikijpanichkul, Planning the location of intermodal freight hubs: an agent based approach, in *Proceedings of the IEEE International Conference on Networking, Sensing and Control*, pp. 187–192. IEEE (2007)
120. D. Webster, R. Reed, A material handling system selection model. AIIE Trans. **3**(1), 13–20 (1971)
121. Y. Wei, Y. Fan, X. Weixuan, An integrated methodology for decision making of mining method selection. Int. J. Manuf. Technol. Manage. **5**(1/2), 10–19 (2003)
122. W. Wiesemann, D. Kuhn, B. Rustem, Maximizing the net present value of a project under uncertainty. Eur. J. Oper. Res. **202**(2), 356–367 (2010)
123. L. Xinchun, Z. Weicheng, Z. Youdi, Application of genetic algorithm to optimize the number and size of equipment of surface mine, in *Mine Planning and Equipment Delection 2004: Proceedings of the Thirteenth International Symposium on Mine Planning and Equipment Selection, Wroclaw, Poland, 1-3 September* eds. by M. Hardygora, G. Paszkowska, M. Sikora, pp. 637–639, London (2004). A.A. Balkema
124. R. Zhang, W.Y. Yun, H. Kopfer, Heuristic-based truck scheduling for inland container transportation. OR Spectrum **32**(3), 787–808 (2010)
125. Y. Zhang, *Advances in LTL Load Plan Design*. Ph.D. thesis, Georgia Institute of Technology (2010)
126. L. Zhongzhou, L. Qining, Erlangian cyclic queueing model for shovel-truck haulage systems. in *Mine Planning and Equipment Selection* ed. by R. Singhal (1988), pp. 423–428

Chapter 4
Match Factor Extensions

Christina N. Burt and Louis Caccetta

4.1 Introduction

The mining and construction industries have long held interests in determining the productivity or efficiency of a selected fleet of trucks and loaders. One way to study the efficiency of a fleet is to weigh the efficiencies of the truck fleet and loader fleet against one another. The *match factor* is the ratio of truck arrival times to loader service rates.

The aforementioned industries have used the match factor for many decades as an indicator of productivity performance. As defined by [5], the match factor ratio, $MF_{i,i'}$, for trucks of type i working with loaders of type i' is given as

$$MF_{i,i'} = \frac{t_{i,i'}x_i}{\bar{t}_X x_{i'}}, \tag{4.1}$$

where x_i is the number of trucks of type i, $x_{i'}$ is the number of loaders of type i', $t_{i,i'}$ is the time taken to load truck type i with loader type i', and $\bar{t}_X$ is the average cycle time for all trucks.

This ratio has hitherto relied on the assumption that the truck and loader fleets are homogeneous. That is, all the trucks are of the same type, and all the loaders are of

---

The original version of this chapter was revised: Originator type has been changed and the chapter author names have been included. The erratum to this chapter is available at https://doi.org/10.1007/978-3-319-76255-5_11

C. N. Burt (✉)
Department of Mathematics and Statistics, The University of Melbourne, Parkville, VIC, Australia
e-mail: c.burt@ms.unimelb.edu.au

L. Caccetta
Department of Mathematics and Statistics, Curtin University, Bentley, WA, Australia
e-mail: L.Caccetta@curtin.edu.au

C. N. Burt and L. Caccetta (eds.), *Equipment Selection for Mining: With Case Studies*, Studies in Systems, Decision and Control 150,
https://doi.org/10.1007/978-3-319-76255-5_4

the same type. In reality, mixed fleets are common. Heterogeneous fleets can occur when equipment types are discontinued, equipment is superseded, or simply when a mixed fleet is cheaper than a homogeneous fleet. Heterogeneous fleets can occur when new equipment is purchased to work alongside pre-existing equipment. It is also possible that a heterogeneous fleet can represent a minimum cost equipment selection solution. In this chapter we propose new ways for defining match factor for heterogeneous fleets.

In particular, we:

- Present two ways to define match factor when heterogeneous truck fleets are present [Sect. 4.2];
- Present a new method to define match factor when heterogeneous loading fleets are operating [Sect. 4.3]; and
- Present a new method to define match factor when both truck and loader fleets are heterogeneous [Sect. 4.4].

The aim of this chapter is to provide extensions to the productivity and efficiency measures currently available in the literature. This will enable greater consideration of heterogeneous fleets.

4.2 Heterogeneous Truck Fleets

The fleet most likely to be heterogeneous is the trucking fleet. This is due to the large number of trucks required to meet production requirements, compared to a relatively small number of loaders. Also, although there may often be different types of loaders operating in a mine, they are often in different locations and so can't be considered as separate fleets.

We begin by considering the truck arrival rate in the case of a heterogeneous truck fleet with a homogeneous loading fleet.

Definition 4.2.1 The **truck arrival rate**, A, for a heterogeneous truck fleet with homogeneous loading fleet is the ratio of the number of trucks to the truck cycle time:

$$A = \frac{\sum_i x_i}{\bar{t}_X},$$

where x_i is the number of trucks of type $i \in \mathbf{X}$ (where $\mathbf{X}$ is the set of all truck types), and $\bar{t}_X$ is the average cycle time for all truck types.

At this stage this rate is unaffected by the number of truck types, as we use an average truck cycle time. The loader service rate is the number of trucks that are served per second. The loader cycle time may vary between different truck types.

Definition 4.2.2 The **loader service rate**, $D_{i'}$, for loader type i' is given by

$$D_{i'} = \frac{x_{i'} \sum_i x_i}{\sum_i t_{i,i'} x_i},$$

where x_i is the number of trucks of type i, $x_{i'}$ is the number of loaders of type $i' \in \mathbf{X}'$, and $t_{i,i'}$ is the time required to load truck type i from loader type i'.

As the match factor is the ratio of truck arrival rate to loader service time, the match factor for heterogeneous truck fleets is easily derived from Definitions 4.2.1 and 4.2.2:

$$MF_{i'} = \frac{A}{D_{i'}} \tag{4.2}$$

$$= \frac{\sum_i x_i}{\bar{t}_X} \Big/ \frac{\sum_i x_i x_{i'}}{\sum_i t_{i,i'} x_i} \tag{4.3}$$

$$= \frac{\sum_i t_{i,i'} x_i}{\bar{t}_X x_{i'}} \tag{4.4}$$

$$= \sum_i MF_{i,i'} \,. \tag{4.5}$$

It is clear that if only one truck type is operating in the fleet, then Eq. (4.4) will produce the same result as Eq. (4.1). Another way to think of this match factor for heterogeneous truck fleets is to add the individual match factors from each of the homogeneous sub-fleets. Note that the alternative method is only appropriate for the case of homogeneous loader fleets working with heterogeneous trucking fleets.

Sometimes we would like to use unique truck cycle times for different truck types in the fleet. This can occur when trucks have different routes. For example, consider a case where larger equipment is used to haul waste while smaller trucks are used to haul ore: the waste and ore may be sent to different locations, with significantly different cycle lengths. When individual truck cycle times are used, the times must be weight averaged to produce an accurate match factor. Equation (4.4) can be easily extended to account for unique truck cycle times.

Definition 4.2.3 The **average cycle time**, $\bar{t}_X$, is given by:

$$\bar{t}_X = \frac{\sum_i t_i x_i}{\sum_i x_i},$$

where t_i is the cycle time for truck type $i \in \mathbf{X}$ and x_i is the number of trucks of type i.

Now, substituting this new truck cycle time into Eq. (4.4), we have the following lemma:

Lemma 4.1 *For heterogeneous truck fleets with individual truck cycle times, the* ***match factor*** *for homogeneous loader fleets of type* $i' \in \mathbf{X}'$ *can be represented by*

$$MF_{i'} = \frac{\left(\sum_i x_i\right)\left(\sum_i t_{i,i'} x_i\right)}{x_{i'} \sum_i t_i x_i}. \tag{4.6}$$

4.3 Heterogeneous Loader Fleets

This section considers the case of mixed loaders in the fleet, while the trucks remain uniform in type. The time required to load a truck may be different for various types of loaders. The loader service rate is the number of trucks served in a defined time period. In a heterogeneous fleet, the time taken to serve a truck may differ between the varying loader types.

Lemma 4.2 *The* ***loader service rate*** *for heterogeneous loader fleets working with truck type* $i \in \mathbf{X}$ *is given by*

$$D_i = \sum_{i'} \frac{x_{i'}}{t_{i,i'}}. \tag{4.7}$$

Proof The loader service rate is the ratio of the total number of trucks to the time required to serve them. We have $t_{i,i'}$ for several loader types i' and one truck type i. The number of trucks type i served by loader type i' in $t_{i,i'}$ time is:

$$\frac{1}{t_{i,i'}}.$$

Thus the total number of trucks served by all loader types in a unit of time is

$$D_i = \sum_{i'} \frac{x_{i'}}{t_{i,i'}}, \text{ as required.}$$

■

Recall that the match factor is the ratio of truck arrival rate to loader service rate. The truck arrival rate is:

$$A_i = \frac{x_i}{t_i},$$

which gives the following theorem:

Theorem 4.1 *For heterogeneous loader fleets, the **match factor** for a homogeneous truck fleet of type $i \in \mathbf{X}$ is*

$$MF_i = \frac{x_i}{t_i \sum_{i'} \frac{x_{i'}}{t_{i,i'}}}. \tag{4.8}$$

When only one type of loader operates in the fleet, Eq. (4.8) reduces to Eq. (4.1). In the case of multiple dump locations or routes, Eq. (4.8) can be expanded to account for differing truck cycle times. First, we represent the average truck cycle time by:

$$\bar{t}_X = \frac{\sum_h t_{i,h} x_{i,h}}{x_i} \tag{4.9}$$

where $t_{i,h}$ is the cycle time for truck type i on route h, and $x_{i,h}$ is the number of trucks of type i working on route h. This gives the following corollary:

Corollary 4.2 *The **match factor** for heterogeneous loader fleets working with truck type i (with individual truck cycle times for trucks on route h) can be represented by*

$$MF_i = \frac{(x_i)^2}{\left(\sum_{i'} \frac{x_{i'}}{t_{i,i'}}\right)\left(\sum_h t_{i,h} x_{i,h}\right)}. \tag{4.10}$$

4.3.1 Example

The following example calculates the match factor of a heterogeneous loader fleet. Table 4.1 outlines the equipment set.

The cycle time for the loader is the time taken for one full swing of the bucket. Some trucks may need several buckets to fill their trays. The first step is to determine the unique loading time for each truck. If the truck capacity is not a round multiple of the loader capacity, then we take another scoop if the capacity left is more than one third of the loader bucket size. This is because it takes almost the same amount of time to move a portion of a scoop as it does to move a full scoop [4].

Truck type A and loader type B: $\frac{150}{60} = 2.5$, and so 3 swings are needed: $3 \times 35 = 105$ seconds

Truck type A and loader type C: $\frac{150}{42} = 3.6$, and so 4 swings are needed: $4 \times 35 = 140$ seconds

Table 4.1 Example data for a heterogeneous loader fleet with common truck cycle time

	Equipment	Capacity (Tonnes)	Cycle Time (s)
22	Truck type A	150	1500
1	Loader type B	60	35
1	Loader type C	42	35

This gives the match factor:

$$\begin{aligned} MF_A &= \frac{x_A}{\bar{t}_X \sum_{i'} \frac{x_{i'}}{t_{A,i'}}} \\ &= \frac{22}{1500 \times (\frac{1}{105} + \frac{1}{140})} \\ &= 0.88. \end{aligned}$$

This shows that the fleet is under-trucked. When a minimum cost fleet is desired, one would reasonably expect that under-trucking would provide better solutions than perfectly matching the fleets with a match factor of 1.

4.4 Heterogeneous Truck and Loader Fleets

For heterogeneous truck and loader fleets, we consider the time required for each loader to serve the available truck fleet. This is equal to the sum of the number of trucks of type i multiplied by the time required to serve that truck type. We call this the *loading times*, $t_{i'}$, for each loader type i'.

Definition 4.4.1 The time, $t_{i'}$, required for a loader of type i' to serve the entire fleet of trucks is

$$t_{i'} = \sum_i t_{i,i'} x_i \,. \tag{4.11}$$

So the time taken for one loader to serve one truck is:

$$\frac{t_{i'}}{\sum_i x_i}. \tag{4.12}$$

Thus we obtain the following definition:

Definition 4.4.2 The **loader service rate** for heterogeneous trucks and loaders is given by

$$D = \sum_{i'} \frac{x_{i'} \sum_i x_i}{t_{i'}} \,. \tag{4.13}$$

As in Sect. 4.3, the truck cycle time is assumed to be constant for the entire truck fleet for that period.

Theorem 4.3 *The* ***match factor*** *for both heterogeneous truck and loader fleets can be represented by*

$$MF = \left(\bar{t}_X \sum_{i'} \frac{x_{i'}}{t_{i'}} \right)^{-1} . \tag{4.14}$$

Proof We now consider the truck arrival rate for the entire fleet, given in Definition 4.2.1:

$$\begin{aligned} MF &= \frac{A}{D} \\ &= \frac{\sum_i x_i}{\bar{t}_X} \times \frac{1}{\sum_i x_i \sum_{i'} \frac{x_{i'}}{t_{i'}}} \\ &= \frac{1}{\bar{t}_X \sum_{i'} \frac{x_{i'}}{t_{i'}}}, \text{ as required.} \end{aligned}$$

■

If we have unique truck cycle times, Eq. (4.14) can be easily extended to:

$$MF = \frac{1}{\left(\sum_{i'} \frac{x_{i'}}{t_{i'}} \right)} \sum_{i \in I} \frac{x_i}{(t_{i,I} x_{i,I})} \tag{4.15}$$

where $t_{i,I}$ is the unique truck cycle time for trucks in subset $I \in \mathbf{X}$.

We find an expression that is equivalent to Eq. (4.14) but may be simpler to implement in a spreadsheet by first observing that the loader service rate can be represented by the following expression (where $\prod$ denotes *product*):

$$D = \frac{\sum_{i'} x_{i'} \sum_i x_i \prod_{h \neq i'} (t_{i,h} x_i)}{\prod_{i,i'} (t_{i,h} x_i)} .$$

We take the truck arrival rate for the entire fleet from Definition 4.2.1 to obtain the following theorem.

Theorem 4.4 *The match factor for both heterogeneous truck and loader fleets can be represented by*

$$MF = \frac{\prod_{i,i'} t_{i,i'} x_i}{\prod_{h \neq i'} (t_{i,h} x_i) \sum_{i'} x_{i'} \bar{t}_X} . \tag{4.16}$$

When only one type of truck and one type of loader operate in the fleet, Eqs. (4.14)–(4.15) reduce to Eq. (4.1), as expected.

Table 4.2 Example data for a heterogeneous truck and loader fleet with common truck cycle time

	Equipment	Capacity (Tonnes)	Cycle Time (s)
15	Truck type A	150	1500
7	Truck type B	230	1500
1	Loader type C	60	35
1	Loader type D	38	30

4.4.1 *Example*

This example determines the match factor of a heterogeneous truck and loader fleet. Table 4.2 presents the data set.

The unique loading times for each truck are determined by the rule of thumb described in Sect. 4.3.

Truck type A and loader type C: $\frac{150}{60} = 2.5$, and so 3 swings are needed: $3 \times 35 = 105$ seconds
Truck type A and loader type D: $\frac{150}{38} = 3.9$, and so 4 swings are needed: $4 \times 30 = 120$ seconds
Truck type B and loader type C : $\frac{230}{60} = 3.8$, and so 4 swings are needed: $4 \times 35 = 140$ seconds
Truck type B and loader type D: $\frac{230}{38} = 6.1$, and so 6 swings are needed: $6 \times 30 = 180$ seconds

We calculate the *loading times*, $t_{i'}$, for each loader of type i'.

$$t_C = 15 \times 105 + 7 \times 140 = 2555$$
$$t_D = 15 \times 120 + 7 \times 180 = 3060$$

$$MF = \frac{1}{\left[\frac{1}{2555} + \frac{1}{3060}\right] \times 1500} = 0.928$$

This solution is close to the theoretical perfect match of 1.0. This is a good result in terms of overall efficiency and productivity of the fleet. However, one should be aware that costing has not been considered in determining the match factor and so it is possible that the fleet would be cheaper to operate even if the match factor was lower.

4.5 Conclusion

For the mining industry, the match factor ratio is an important performance indicator which we have extended for several likely circumstances, including heterogeneous truck and loader fleets with multiple routes. The match factor can be used to optimise the truck cycle time in order to gain maximal efficiency from the selected fleet. Alternatively, project managers may use the match factor formula to determine the ideal number of trucks in the fleet. The formulae presented in this chapter are less restricted in their choice of equipment, select mixed fleets to suit the productivity requirements and minimise materials handling expense.

It is interesting to note that the match factor ratio in [5] excludes waiting and queuing times for trucks and loaders. This may be because the waiting time for a truck fleet is difficult to estimate without first knowing the size of the truck fleet. However, if we use the match factor ratio as an index of overall fleet efficiency, then it is acceptable to include waiting times that have been estimated by other methods.

The formulae presented in this chapter provide a sensible extension to the original formula and bring greater accuracy to the cases where mixed fleets operate together. All of these formulae can be implemented easily in spreadsheet software such as Microsoft Excel. Throughout the rest of this book, we employ the heterogeneous match factor ratios to indicate the overall efficiency of the selected fleets.

Acknowledgements The contents of this chapter appear in [1–3].

References

1. C. Burt, An optimisation approach to materials handling in surface mines. Ph.D. thesis, Mathematics and Statistics, Curtin University of Technology, Bentley, Perth, Australia, 2008, http://espace.library.curtin.edu.au:80/R?func=dbin-jump-full&local_base=gen01-era02&object_id=165737
2. C.N. Burt, L. Caccetta, Match factor for heterogeneous truck and loader fleets. Int. J. Min. Reclam. Environ. **21**, 262–270 (2007)
3. C.N. Burt, L. Caccetta, Corrigendum: match factor for heterogeneous truck and loader fleets. Int. J. Min. Reclam. Environ. **22**(1), 84–85 (2008)
4. D.D. Gransberg, Optimizing haul unit size and number based on loading facility characteristics. J. Constr. Eng. Manag. **122**(3), 248–253 (1996)
5. W.C. Morgan, L.L. Peterson, Determining shovel-truck productivity. Min. Eng. 76–80 (1968)

Part II
Optimisation Models and Case Studies

Chapter 5
Case Studies

Christina N. Burt and Louis Caccetta

5.1 Introduction

In this chapter, we provide the data and background for two case studies. The aim is to provide data for mine scenarios of varied complexity. For example, in our first case study, we begin with a simple mine set-up, with interesting variation in truck-cycle time. In the second case study, we present a mine with many mining locations, including stockpiles. This data was provided by an in-house equipment selection expert from an industry partner. Some of the data is publicly available, such as the equipment size, swing-time, travel speeds, etc.; while other data is mine-specific, such as truck-cycle times for each period and expected production.

We have anonymised the equipment in the data-sets in order to protect contractual agreements between our industry partner and their equipment suppliers. We also anonymise the mines whose data is described, to protect the interests of our industry partner. We feel that these two anonymisation steps do not detract from the data-sets overall, and still allow for meaningful analysis.

The very nature of data in the mining industry is approximate. From the beginnings, the design of the pit is based on ore-body estimations which are often inaccurate and difficult to validate due to the expense of accurate sampling (assaying). This can

The original version of this chapter was revised: Originator type has been changed and the chapter author names have been included. The erratum to this chapter is available at https://doi.org/10.1007/978-3-319-76255-5_11

C. N. Burt (✉)
Department of Mathematics and Statistics, The University of Melbourne, Parkville, VIC, Australia
e-mail: c.burt@ms.unimelb.edu.au

L. Caccetta
Department of Mathematics and Statistics, Curtin University, Bentley, WA, Australia
e-mail: L.Caccetta@curtin.edu.au

C. N. Burt and L. Caccetta (eds.), *Equipment Selection for Mining: With Case Studies*, Studies in Systems, Decision and Control 150,
https://doi.org/10.1007/978-3-319-76255-5_5

mean that from the beginning, the expected production figures for each period can be very different by the time we reach the period in question. Some data has been pre-filtered in a way that lessens its accuracy, such as the cost of equipment performance over time—this is typically discretised, and it is challenging or impossible to find the original, nonlinear performance functions. Thus all data and results should be taken in this context, where the accumulation of approximations may lead to a substantially different solution that what may be achieved with the 'real' data, were it available. However, the data at hand still enables analysis of the system and the design of a robust model to solve the data for the case where more accurate data may become available.

We begin the chapter with a 9-period, few-locations case study in Sect. 5.2, where we describe the locations and routes, production requirements and case-specific parameters. We follow suit with the 13-period, many-locations case study in Sect. 5.3. We conclude with a section on compatibility and availability of equipment, which is relevant for both case studies.

5.2 Few-Locations Case Study

Our industry partner wishes to select a fleet of trucks and loaders for an open pit iron ore mine operating under a truck-loader hauling system. This mine is in the planning stages and begins with no pre-existing equipment.

5.2.1 Locations and Routes

For this mine, there are only two mining locations. Each location produces ore which needs to be delivered to the mill, and waste which needs to be delivered to a dump site. Figure 5.1 describes the two mining locations and four routes which connect them to the waste dump site and the mill.

5.2.2 Production Requirements

This case study considers a mine operating under a truck-loader hauling system and mines ore and waste in an open pit. Our industry partner provided the production requirement data for both the mining locations and the truck routes. For this case study, we consider $K = 9$ periods in total, each of length 1 year. This new mine is simple in terms of the number of mining locations. The overall production requirements change drastically as the overburden is removed, and as the routes become longer (Table 5.1). For example, in period one, the production requirements are only 2.13 million tonnes. This grows to around 19 million tonnes for the subsequent 3 years. The estimated truck cycle times (also provided by the industry partner) also demon-

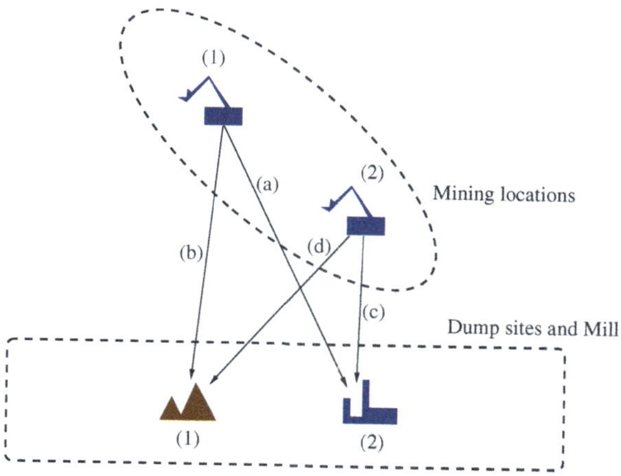

Fig. 5.1 The locations for case study one: the few-locations case study

strates great variability from period to period, and between locations (Table 5.2). For example, the smallest truck cycle time is 2.64 m, while the longest is 22.82 m.

5.2.3 *Case Specific Parameters*

We have the following parameters supplied by our industry partner:

- The mine is removing ore and waste, and operates under a shovel-truck system.
- The mine operates for 7604 h in each period (accounting for blasting days, holidays and other non-operational days).
- The loaders are selected from a set of 20 loader types.
- The trucks are selected from a set of 8 truck types.
- There are K = 9 periods in total, each of length 1 year.
- The cost-bracket length, B_0, is 5000 h.
- The interest rate for all periods is 8%.
- The depreciation rate is set to 50%.
- The maximum value for any truck variable is 30, i.e., maximum 30 trucks of a given type, in any age bracket, in any period.
- The maximum value for any loader variable is 10, i.e., maximum 10 loaders of a given type, in any age bracket, in any period.

Note that the depreciation value is used to estimate the sale (or salvage) value of used equipment, rather than for tax-offset purposes. We choose a high depreciation value to lessen the impact of a sale on the decisions, as the second hand market is unreliable.

Our industry partner also provided Utilisation Factors, which are reducing factors to account for lost hours due to inefficiency, maintenance, and availability of

Table 5.1 The production requirements for the routes for the few-locations case study

Route	Period								
	1	2	3	4	5	6	7	8	9
(a)	51701	668713	1602777	1463289	2227402	1657740	2031283	2230192	2474503
(b)	2082800	8843227	7294449	5837564	5356417	8051374	4582733	4666783	2001808
(c)	0	0	1189184	1120125	337101	1073245	1106596	455511	235810
(d)	0	9412391	9063476	10593225	11355932	4091566	530276	84354	94315
Total (MT)	2.13	18.92	19.14	19.01	19.27	14.87	8.25	7.43	4.80

Key for routes
(a) Mining location (1) to Mill
(b) Mining location (1) to Dump site (1)
(c) Mining location (2) to Mill
(d) Mining location (2) to Dump site (1)

Table 5.2 The truck cycle times (minutes) for the few-locations case study

Period	Route			
	(a)	(b)	(c)	(d)
1	8.24	2.64	0.0	0.0
2	8.3	3.48	0.0	8.24
3	9.28	3.84	5.74	10.23
4	10.52	4.88	8.73	10.45
5	11.16	6.01	10.38	12.6
6	12.47	7.23	11.71	16.72
7	12.05	8.49	13.82	19.6
8	15.77	10.11	15.49	21.37
9	17.74	12.05	16.52	22.82

the equipment; and, a compatibility matrix between all equipment types. These are provided in later sections, as they are relevant to both cases.

5.3 Many-Locations Case Study

Our industry partner provided a second case study from an ongoing mining operation with pre-existing equipment and a more complex route structure. This case study considers a surface mine operating under a truck-loader hauling system, and mines ore and waste in an open pit.

5.3.1 Locations and Routes

The mine for this case study has eight loading locations—four mining locations and four stockpiles, as depicted in Fig. 5.2. Mixing constraints are the quantity of different grades of ore required to make up the final grade. Typically the final grade is determined by market demand. The mixing constraints are not considered in this model, as they are assumed to be pre-optimized when the mine plan is produced. In this case study the stockpiles are old, and newly mined ore or waste is not dumped in these locations. Instead, they are used to create the appropriate mix at the mill—so the flow from these locations is unidirectional. There are also four dump sites including one mill. Connecting these mining locations, stockpiles and dump site, are 13 routes in total (route key provided in Table 5.3).

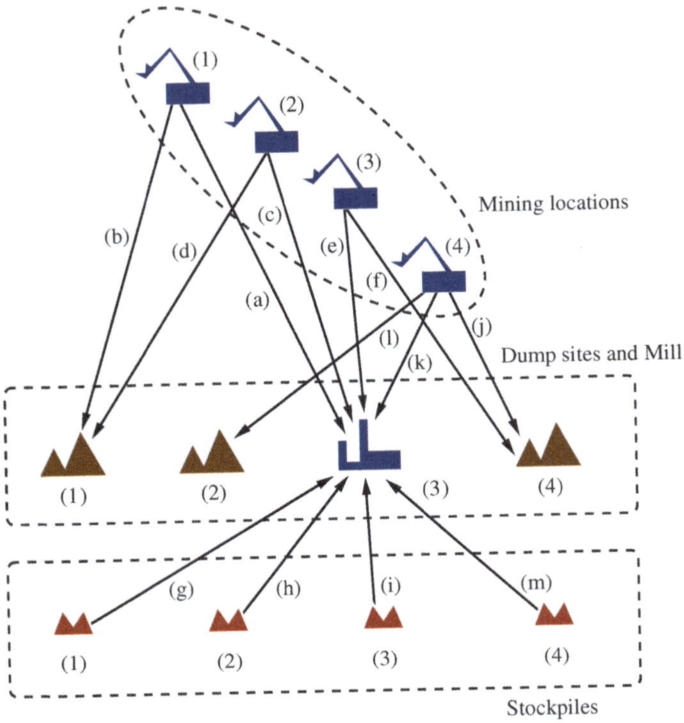

Fig. 5.2 Routes from mining locations to dump sites for the many-locations case study

5.3.2 *Production Requirements*

Our industry partner provided the production requirement data for both the mining locations and the truck routes. Tables 5.3 and 5.4 describe the quantity of material (tonnes) to be moved from each location, to each dump site. Our industry partner also provided pre-estimated truck cycle times for each route, presented in Table 5.5. We combined the location data to create a production requirement graph for the mine over all periods in Fig. 5.3.

5.3.3 *Pre-existing Equipment*

We begin the mine plan with some pre-existing equipment, as listed in Table 5.6. This includes eleven 172 tonne trucks of varying age in hours; and three loaders, namely two 34 tonne hydraulic shovels and one 42 tonne hydraulic shovel.

Table 5.3 The production requirements (tonnes) for the truck routes for 13-period, many-locations case study. Dump site (3) is the mill—we treat it as a regular dump site

Route	Period												
	1	2	3	4	5	6	7	8	9	10	11	12	13
(a)	2598	0	0	0	0	0	0	0	0	0	0	0	0
(b)	455	0	0	0	0	0	0	0	0	0	0	0	0
(c)	203	10141	7659	0	0	0	0	0	0	0	0	0	0
(d)	741	7060	2964	0	0	0	0	0	0	0	0	0	0
(e)	221	827	5928	12797	9919	0	0	0	13990	13990	13990	5184	11
(f)	2592	22935	41035	26948	19873	0	0	0	8901	18191	12589	5572	32
(g)	0	650	0	0	0	0	0	0	0	0	0	0	0
(h)	0	270	809	0	0	0	0	0	0	0	0	0	0
(i)	0	1142	0	0	0	0	0	0	0	0	0	0	0
(j)	0	0	0	737	4071	12990	12890	13990	0	0	0	7713	0
(k)	0	0	0	17051	24126	8984	8583	12836	0	0	0	6251	0
(l)	0	0	0	0	0	799	0	0	0	0	0	0	0
(m)	0	0	0	0	0	0	798	0	0	0	0	0	0

Key for routes
(a) Mining location (1) to Dump site (3)
(b) Mining location (1) to Dump site (1)
(c) Mining location (2) to Dump site (3)
(d) Mining location (2) to Dump site (1)
(e) Mining location (3) to Dump site (3)
(f) Mining location (3) to Dump site (4)
(g) Stockpile (1) to Dump site (3)
(h) Stockpile (2) to Dump site (3)
(i) Stockpile (3) to Dump site (3)
(j) Mining location (4) to Dump site (1)
(k) Mining location (4) to Dump site (2)
(l) Mining location (4) to Dump site (3)
(m) Stockpile (4) to Dump site (3)

Table 5.4 The production requirements (tonnes) for the mining locations for 13-period, many-locations case study

Location	Period												
	1	2	3	4	5	6	7	8	9	10	11	12	13
M(1)	3053	0	0	0	0	0	0	0	0	0	0	0	0
M(2)	944	17201	10624	0	0	0	0	0	0	0	0	0	0
M(3)	2813	23762	46963	39746	29792	0	0	0	22891	32181	26579	10756	43
M(4)	0	650	0	0	0	0	0	0	0	0	0	0	0
S(1)	0	270	809	0	0	0	0	0	0	0	0	0	0
S(2)	0	1142	0	0	0	0	0	0	0	0	0	0	0
S(3)	0	0	0	17789	28197	22773	21474	26826	0	0	0	13964	0
S(4)	0	0	0	0	0	0	798	0	0	0	0	0	0

Key for locations
M(i) Mining location i
S(j) Stockpile j

Table 5.5 Truck cycle times for the 13-period, many-locations case study

Period	Routes												
	(a)	(b)	(c)	(d)	(e)	(f)	(g)	(h)	(i)	(j)	(k)	(l)	(m)
1	35.56	35.17	38.49	37.89	11.23	10.62	0.0	0.0	0.0	0.0	0.0	0.0	0.0
2	0.0	0.0	38.03	37.42	14.97	17.23	16.05	60.00	20.00	0.0	0.0	0.0	0.0
3	0.0	0.0	40.09	40.77	25.02	22.08	0.0	60.00	0.0	0.0	0.0	0.0	0.0
4	0.0	0.0	0.0	0.0	27.18	25.09	0.0	0.0	0.0	34.39	34.27	0.0	0.0
5	0.0	0.0	0.0	0.0	28.58	26.25	0.0	0.0	0.0	26.57	27.38	0.0	0.0
6	0.0	0.0	0.0	0.0	0.0	0.0	0.0	0.0	0.0	32.20	33.51	37.43	0.0
7	0.0	0.0	0.0	0.0	0.0	0.0	0.0	0.0	0.0	34.48	35.71	0.0	15.00
8	0.0	0.0	0.0	0.0	0.0	0.0	0.0	0.0	0.0	37.66	38.76	0.0	0.0
9	0.0	0.0	0.0	0.0	32.64	29.41	0.0	0.0	0.0	0.0	0.0	0.0	0.0
10	0.0	0.0	0.0	0.0	36.42	32.47	0.0	0.0	0.0	0.0	0.0	0.0	0.0
11	0.0	0.0	0.0	0.0	38.53	36.49	0.0	0.0	0.0	0.0	0.0	0.0	0.0
12	0.0	0.0	0.0	0.0	43.54	43.57	0.0	0.0	0.0	39.52	46.67	0.0	0.0
13	0.0	0.0	0.0	0.0	46.93	45.00	0.0	0.0	0.0	0.0	0.0	0.0	0.0

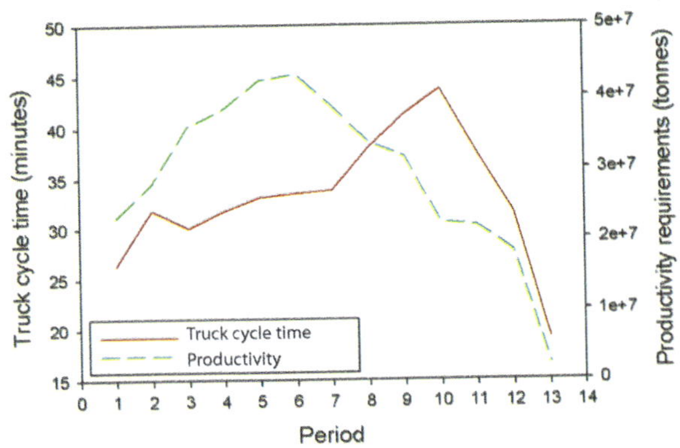

Fig. 5.3 Production requirements and truck cycle times for the 13-period, many-locations case study

Table 5.6 Pre-existing equipment for case study two

Equipment i.d.	L_7^P	L_7^P	L_{17}^P	T_{12}^P	T_{12}^P	T_{12}^P
Quantity	1	1	1	3	2	6
Capacity (tonnes)	34	34	42	172	172	172
Age (years)	16	17	16	7	8	11

5.3.4 *Case Specific Parameters*

Some parameters are defined by the industry partner:

- The mine operates for 7604 h in each period (accounting for blasting days, holidays and other non-operational days);
- The loaders are selected from a set of 20 loader types;
- The trucks are selected from a set of 8 truck types;
- The cost-bracket partition, B_0, is 5000 h;

Table 5.7 Compatibility of trucks with loaders

Truck type	Loader type																										
	0	1	2	3	4	5	6	7	8	9	10	11	12	13	14	15	16	17	18	19	20	21	22	23	24	25	26
0	1	1	1	1	1	1	1	1	1	1	1	1	1	1	1	1	1	1	1	1	1	1	1	1	1	1	1
1	1	0	0	0	1	1	0	0	0	0	1	0	1	0	1	0	1	0	1	0	1	0	1	0	0	1	0
2	0	1	0	0	0	1	0	0	1	0	1	0	1	0	1	0	1	0	1	0	1	1	1	0	0	1	0
3	1	1	1	1	1	1	0	1	1	1	1	1	1	1	1	1	1	1	1	1	1	1	1	1	1	1	1
4	1	1	1	0	0	1	0	1	1	1	1	1	1	0	1	0	1	1	1	1	1	1	1	1	1	1	0
5	0	1	1	0	1	1	0	1	1	1	1	1	1	0	1	0	1	1	1	0	1	1	1	1	1	0	0
6	1	1	1	1	0	0	0	1	1	1	1	1	1	1	1	1	1	1	1	1	0	1	1	1	1	1	1
7	1	1	1	1	0	0	0	1	1	1	1	1	1	1	1	1	1	1	1	1	0	1	1	1	1	1	1
8	1	1	1	1	1	0	0	1	1	1	1	1	1	1	1	1	1	1	1	1	1	1	1	1	1	1	1
9	1	1	1	1	1	1	1	1	1	1	1	1	1	1	1	1	1	1	1	1	1	1	1	1	1	1	1
10	1	1	1	1	0	0	0	1	1	1	1	1	1	1	1	1	1	1	1	1	0	1	1	1	1	1	1
11	1	1	1	1	1	0	0	1	1	1	1	1	1	1	1	1	1	1	1	1	1	1	1	1	1	1	1
12	1	1	1	1	1	0	0	1	1	1	1	1	1	1	1	1	1	1	1	1	1	1	1	1	1	1	1
13	1	1	1	1	1	0	0	1	1	1	1	1	1	1	1	1	1	1	1	1	1	1	1	1	1	1	1
14	1	1	1	1	1	1	1	1	1	1	1	1	1	1	1	1	1	1	1	1	1	1	1	1	1	1	1
15	1	1	1	1	1	1	0	1	1	1	1	1	1	1	1	1	1	1	1	1	0	1	1	1	1	1	1
16	1	1	1	1	1	1	0	1	1	1	1	1	1	1	1	1	1	1	1	1	1	1	1	1	1	1	1
17	1	1	1	1	0	1	0	1	1	1	1	1	1	1	1	1	1	1	1	1	1	1	1	1	1	1	1
18	1	1	1	1	0	0	0	1	1	1	1	1	1	1	1	1	1	1	1	1	1	1	1	1	1	1	1
19	1	1	1	1	1	0	1	1	1	1	1	1	1	1	1	1	1	1	1	1	1	1	1	1	1	1	1
20	1	1	1	1	1	0	0	1	1	1	1	1	1	0	1	0	1	1	1	0	0	1	1	1	1	0	0
21	0	1	1	0	1	0	0	1	1	1	1	1	1	0	1	0	1	1	1	0	0	1	1	1	1	0	0
22	1	1	1	1	0	0	0	1	1	1	1	1	1	1	1	1	1	1	1	1	0	1	1	1	1	1	1
23	1	1	1	1	0	0	0	1	1	1	1	1	1	1	1	1	1	1	1	1	0	1	1	1	1	1	1

Table 5.8 The availability of trucks and loaders for each 5000 h age bracket

Age bracket											
	1	2	3	4	5	6	7	8	9	10	
Electric rope shovel	0.9	0.9	0.9	0.9	0.89	0.88	0.87	0.86	0.85	0.84	
Hydraulic shovel	0.9	0.9	0.89	0.88	0.87	0.86	0.85	0.84	0.82	0.8	
Front end loader	0.9	0.88	0.86	0.84	0.82	0.85	0.83	0.81	0.79	0.8	
Trucks	0.92	0.92	0.91	0.91	0.9	0.9	0.89	0.89	0.885	0.881	
	11	12	13	14	15	16	17	18	19	20	21
Electric rope shovel	0.83	0.82	0.81	0.8	0.8	0.8	0.8	0.8	0.8	0.8	0.8
Hydraulic shovel	0.78	0.76	0.74	0.72	0.7	0.7	0.7	0.7	0.7	0.7	0.7
Front end loader	0.78	0.76	0.74	0.72	0.7	0.68	0.66	0.64	0.62	0.6	0.58
Trucks	0.877	0.873	0.869	0.865	0.861	0.857	0.853	0.849	0.845	0.841	0.837

- The schedule is 13 years long;
- The discount rate for all periods is 8% (the approximate interest rate obtainable on investments).

We define the following parameters:

- There are 13 periods, K, each of length 1 year;
- The depreciation rate is set to 50% (a rate of 40–60% is common for this application due to unreliability in the second hand equipment market);
- The maximum value for any truck variable is 30, i.e., maximum 30 trucks of a given type, in any age bracket, in any period;
- The maximum value for any loader variable is 10, i.e., maximum 10 loaders of a given type, in any age bracket, in any period.

5.4 Compatibility and Availability

The compatibility matrix is given in Table 5.7 for all equipment types, where a 0 indicates that a truck type is not compatible with a loader, and a 1 indiates that the pair is compatible. Equipment availability over age brackets was provided in Table 5.8. No leased equipment was included in the set. Cost data and equipment identities were provided, but cannot be presented here for confidentiality reasons.

Acknowledgements Components of this chapter were presented in [1–3]

References

1. C. Burt, An optimisation approach to materials handling in surface mines. Ph.D. thesis, Mathematics and Statistics, Curtin University of Technology, Bentley, Perth, Australia, 2008. http://espace.library.curtin.edu.au:80/R?func=dbin-jump-full&local_base=gen01-era02&object_id=165737
2. C. Burt, L. Caccetta, P. Welgama, L. Fouché, Equipment selection with heterogeneous fleets for multiple period schedules. J. Oper. Res. Soc. **62**, 1498–1509 (2011)
3. C. Burt, L. Caccetta, L. Fouché, P. Welgama, An MILP approach to multi-location, multi-period equipment selection for surface mining with case studies. J. Ind. Manage. Optim. **12**(2), 403–430 (2016)

Chapter 6
Single Location Equipment Selection

Christina N. Burt, Louis Caccetta, Palitha Welgama and Leon Fouché

6.1 Introduction

In this chapter, we provide an equipment selection tool that will select the trucks and loaders for a single-location, multi-period mine at minimum cost. A single-location mine is a mine which has just one mining location with a single route to a single dump-site (Fig. 6.1). By limiting the model to a single location, we are generalising the productivity requirements for the mine to just one location. Therefore, this model is a most useful tool for small mines or for mines where there is no significant difference in the mining locations and route lengths. It can also be used as a tool for larger and more complicated mining scenarios where an approximate solution may be sought for a large number of periods in a relatively short computation time.

We wish to consider all time periods in the mining schedule to ensure that the selected fleet can cope with long-term changes, rather than simply satisfying short-term requirements. In addition to this, we allow for pre-existing equipment. Due to the

The original version of this chapter was revised: Originator type has been changed and the chapter author names have been included. The erratum to this chapter is available at https://doi.org/10.1007/978-3-319-76255-5_11

C. N. Burt (✉)
Department of Mathematics and Statistics, The University of Melbourne, Parkville, VIC, Australia
e-mail: c.burt@ms.unimelb.edu.au

L. Caccetta
Department of Mathematics and Statistics, Curtin University, Bentley, WA, Australia
e-mail: L.Caccetta@curtin.edu.au

P. Welgama · L. Fouché
Rio Tinto Iron Ore, Perth, WA, Australia
e-mail: palitha.Welgama@riotinto.com

L. Fouché
e-mail: Leon.Fouche@riotinto.com

C. N. Burt and L. Caccetta (eds.), *Equipment Selection for Mining: With Case Studies*, Studies in Systems, Decision and Control 150,
https://doi.org/10.1007/978-3-319-76255-5_6

Fig. 6.1 A simplified model of a mine with a single location, single dump-site and a route connecting the two

dynamic nature of mines and the subsequent demands placed on them, it is common for a mining schedule to change significantly after only a few years into the schedule. When this occurs, it may be necessary to re-perform the equipment selection. In this case, there would be a considerable amount of pre-existing equipment—some of which may be discontinued or superseded by better equipment. This also means that we may have heterogeneous fleets.

In addition to the purchase of equipment, we also consider the salvage of equipment for two reasons. Firstly, mines often operate for longer than the standard life of a piece of equipment. Thus, we can expect that we will need to retire some of the equipment during the schedule. In particular, pre-existing equipment may already be close to retirement age. Secondly, mining schedules sometimes have significant changes in productivity requirements from period to period. We wish to determine whether it is better to purchase equipment in the short term to satisfy increased requirements, or to hire equipment over the peak periods. Allowing salvage permits us to investigate this.

In this chapter, we develop an integer program with the following features:

- the inclusion of pre-existing equipment and heterogeneous fleets;
- a multi-period mining schedule;
- linear compatibility constraints ensuring satisfaction of productivity requirements;
- costing which is apportioned to the age of the equipment (in line with industry standards).

Many aspects of the presented model, such as the consideration of multiple periods and pre-existing equipment, are novel for the mining industry and ensure that the model is both a new and advanced equipment selection tool.

We formulate the model in Sect. 6.2 by first listing the assumptions (Sect. 6.2.1), then outlining the decision variables (Sect. 6.2.2), and deriving the objective function (Sect. 6.2.3) and the constraints (Sect. 6.2.4). To test the model we consider a mining case study in Sect. 6.3. Finally, we discuss the results and opportunities to extend the work in Sect. 6.4.

6.2 The Model

Trucks and loaders only come in discrete quantities, so it is appropriate to track them using integer variables. A cost objective function can be linearised through careful definition of the variable indices, and productivity constraints are naturally linear.

Therefore the equipment selection problem for surface mines is best expressed as a pure integer program.

6.2.1 Assumptions

The model is, by necessity, an abstraction of the real problem. Consequently we apply some assumptions and conditions to make the model solvable:

- Known mine schedule—An acceptable mine schedule has already been derived, and the mining method has been selected. We start the equipment selection process with a set of trucks and loaders that suit the mine.
- Single mining location—All the loaders and trucks operate as one fleet. That is, all loaders work in the same location, and all the trucks can service any loaders (subject to compatibility).
- Salvage—All equipment is salvageable at the start of each period at some depreciated value of the original capital expense. Any pre-existing equipment may be salvaged at the start of the first period.
- No auxiliary equipment—We do not consider auxiliary equipment, such as wheel loaders and small trucks. Although the cost of running auxiliary equipment may differ according to the overall fleet selection, we consider this cost to be trivial. Note that the cost of auxiliary equipment can be built into the operating cost if necessary.
- Known operating hours—The operating hours of the mine are estimated by taking planned downtime, blasting and weather delays into account.
- Single truck cycle time—The truck cycle time is the time taken for a truck to be loaded with material, travel to the dump-site, dump the material and return to the loader for queueing. Since we assume a single mining location, a single truck cycle time is used for all trucks. The cycle time is constant over a time period and is known for all periods. It accounts for factors that affect the truck performance, such as rimpull, rolling resistance, haul distance and haul grade.
- Heterogeneous fleets—Different types of equipment can work side-by-side.
- Fleet retention—We retain all equipment at the end of the last period.
- Full period utilisation—All equipment is fully utilised for the entire period in which it is owned.

The length of a time period as used in the model can be adjusted to any desired length, but for the purpose of the case study, we set it to one year.

6.2.2 Decision Variables and Notation

Let the set of all truck types be $\mathbf{X}$, and the set of all loader types be $\mathbf{X}'$. We use i and i' to denote a single truck and loader type, respectively. To simplify the notation where expressions apply to both trucks and loaders, we use e to represent a single type (which can be a truck or loader type), where $e \in \mathbf{X} \cup \mathbf{X}'$.

Suppose that there are K periods, M truck types and N loader types with maximum age L. The following model generates at most $2(MKL + NKL) + K(2^N - 1)$ variables.

6.2.2.1 Fleet Purchase and Operating Variables

For this formulation we wish to capture the amount of operating equipment, the types, ages and the periods in which they operate. We adopt three indices to represent type (e), period (k) and age in the number of periods since it was purchased (l). The index e is drawn from the set of all available equipment types, $\mathbf{X} \cup \mathbf{X}'$; the index k is drawn from the set of all periods, $\{1, \ldots, K\}$; and the index l is drawn from the age range of equipment type e (which can vary significantly amongst types). We define:

$\mathbf{x}_e^{k,l}$: number of owned equipment units of type e that are age l at the start of period k.

6.2.2.2 Salvage Variables

We use salvage variables to tell when equipment is salvaged. These variables are defined similarly to the operating variables:

$\mathbf{s}_e^{k,l}$: number of owned equipment units of type e that are age l which are salvaged at the start of period k.

6.2.2.3 Compatibility Indicator Variables

We introduce an indicator variable for the compatibility constraint set. This variable ensures that either the truck fleet matches the loader productivity levels, or the truck fleet matches the productivity requirements of the mine.

$\mathbf{h}_{A'}^k$: variable indictor for satisfaction of either loader productivity or mine productivity in period k for trucks compatible with loader set $A' \subset \mathbf{X}'$.

This variable will adopt a value of 1 if the mine productivity constraint is active, and a value of 0 if the loader productivity constraint is active.

6.2.3 *Objective Function*

In the mining industry, the viability of a mine depends on the efficiency of the equipment and its ability to meet production requirements at the lowest possible cost. Therefore, we consider the objective function as the cost of materials handling. More specifically, we are interested in the *net present value* (NPV) of the cost of materials handling for the life of the mine. We consider the capital expense, the operating expense, and the salvage value of the equipment with respect to a discount rate I.

First, we consider the capital expense, which is a one-off cost incurred by purchase. We represent the fixed cost of purchasing equipment of type e by F_e, and discount this purchase to the present using a discount factor:

$$D_1^k = \frac{1}{(1+I)^k}, \tag{6.1}$$

where k is the period in which the equipment was bought. 0 is the present time.

Note that pre-existing equipment does not incur a capital expense as it was purchased in a period not considered in this optimisation horizon. Thus the total capital expense for a truck or loader of type e is

$$\sum_{e,k} F_e D_1^k \mathbf{x}_e^{k,0}. \tag{6.2}$$

The operating expense is the cost of operating and maintaining the equipment. It takes into account varying maintenance expenses, availability and productivity levels, which are known to vary with the age of the equipment. In the mining industry, the typically nonlinear operating cost is simplified by creating a piece-wise linear function that is divided into age brackets (Fig. 6.2). We use brackets of size $B_0 = 5000$.

The age bracket in which equipment lies at the beginning of a period is given by $b(l) \in \{0, 1, \ldots, s-1\}$, where s is the total number of age brackets. Let H^k be the operating hours for period k and $a_e^{k,l}$ be the availability of equipment type e, aged l.

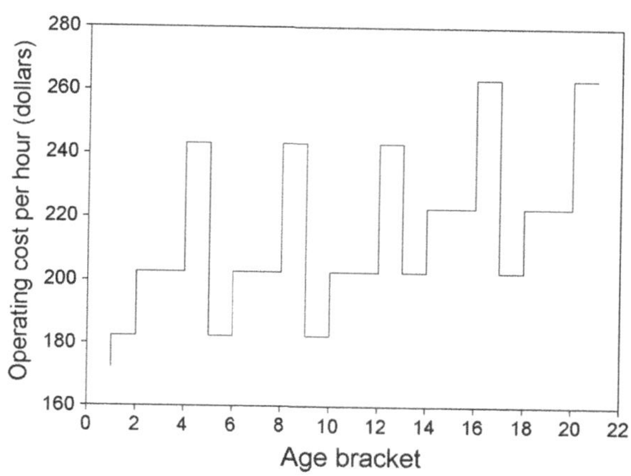

Fig. 6.2 Equipment operating cost versus age bracket. The rise in operating cost reflects the increased maintenance expense over time. Large drops in the operating expense occur when an overhaul has taken place

The availability of equipment is the proportion of the period that the equipment is available to work.

We calculate the age bracket for period k using:

$$b(l) = \left\lfloor \frac{\sum_{h=k-l}^{k-1} a_j^{k,l+n-k} H^n}{B_0} \right\rfloor. \tag{6.3}$$

We assume that the operating cost is constant over an age bracket, so if a piece of equipment moves into a different age bracket within a period, the operating cost for the period should be appropriately split between these two brackets. We wish to determine the proportion of time that the equipment lies in the starting age bracket, and the proportion of time that the equipment lies in the next age bracket (Fig. 6.3).

By assumption, the size of an age bracket exceeds the size of a period, so a piece of equipment cannot occupy more than two age brackets in one period. We represent this proportion of time by the parameter $B_{h,e}^{k,l}$ for $h = 1, 2$, where h is the hth age bracket that the equipment has landed within the period. Then we have:

$$B_{1,e}^{k,l} = \begin{cases} 1 & \text{if } (b(l)+1)B_0 - \sum_{n=k-l}^{k-1} a_j^{k,l+n-k} H^n > a_e^{k,l} H^k, \\ \frac{(b(l)+1)B_0 - \sum_{n=k-l}^{k-1} a_j^{k,l+n-k} H^n}{a_e^{k,l} H^k} & \text{otherwise,} \end{cases}$$

and

$$B_{2,e}^{k,l} = 1 - B_{1,e}^{k,l}.$$

The variable cost, $V_e^{k,b(l)+h-1}$, is the cost per operated hour for equipment type $e \in \mathbf{X} \cup \mathbf{X}'$ in age bracket $b(l) + h - 1$ in period k. Hence, for the operating expense, we have the following expression:

$$\sum_{e,k,l,h} B_{h,e}^{k,l} D_1^k V_e^{k,b(l)+h-1} \mathbf{x}_e^{k,l}. \tag{6.4}$$

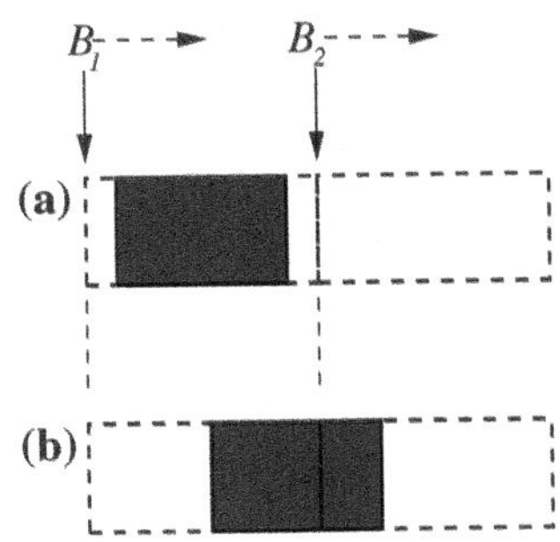

Fig. 6.3 The two cases of equipment age landing between age brackets. For case (**a**), the equipment stays in the same age bracket for the entire period. In case (**b**), the equipment steps over into the next age bracket within the period

As we are minimising the cost of materials handling, we represent salvage by a negative expense. We apply a combined depreciation (at rate J per period) and net present value discount factor (at rate I per period):

$$D_2^{k,l} = \frac{(1-J)^l}{(1+I)^k}, \tag{6.5}$$

where l is the age of the equipment at the start of period k.

Since F_e is the original capital expense, the salvage cost is:

$$-\sum_{e,k,l} F_e D_2^{k,l} \mathbf{s}_e^{k,l}. \tag{6.6}$$

In total there are $3(MKL + NKL) + MK + NK = O((M+N)KL)$ terms in the objective function. The final objective function is the sum of these components and is given in Eq. (6.21).

6.2.4 Constraints

6.2.4.1 Index Restrictions

To reduce the total number of variables in our program, we restrict the age bracket of any piece of equipment to be no greater than the maximum age of the equipment type, which we denote as $L(e)$ for type e. In other words, we do not create a variable for equipment older than its maximum age. Furthermore, we restrict the age of any equipment that is not pre-existing to be no greater than the number of the current time period. This is because equipment cannot increase two age brackets in a single time period. However, we must also take into account the possibility of pre-existing equipment. If we define $P(e)$ to be the highest starting age of any pre-existing equipment of type e, then in time period k we only have to consider equipment up to age

$$L^k(e) = \min\{P(e) + k - 1, L(e)\}.$$

Whenever we sum over l, we only need to sum up to $l = L^k(e)$. We note that as salvage occurs at the start of the period, salvage variables extend to $l = L^k(e) + 1$.

6.2.4.2 Production Constraints

In a mine, the correct quantities of materials must be handled to satisfy the mixing demands of the mill. To capture the production requirements in a constraint, we first consider the potential productivity of the equipment, $P_e^{k,l}$, when it is aged l in period k. We can determine this quantity by looking at the equipment availability (a_e^l), capacity (c_e) and cycle time (t_e^k):

$$P_e^{k,l} = \frac{a_e^{b(l)} c_e}{t_e^k}. \tag{6.7}$$

For this formulation, availability is determined by the equipment's age. The age determines the age bracket under which the equipment falls. In turn, the age bracket determines an availability estimate, which represents the proportion of total time that the equipment is available to work. We have the following production constraints for trucks and loaders respectively, where T^k is the production requirement for period k:

$$\sum_{i,l} P_i^{k,l} \mathbf{x}_i^{k,l} \geq T^k \qquad \forall \quad k, \tag{6.8}$$

$$\sum_{i',l} P_{i'}^{k,l} \mathbf{x}_{i'}^{k,l} \geq T^k \qquad \forall \quad k. \tag{6.9}$$

6.2.4.3 Compatibility Constraints

We must ensure that the trucks and loaders used in a period are compatible with each other. However, we do not need to make all trucks compatible with all loaders; we must merely have enough compatibility to satisfy productivity requirements, not just one set. We define the set $\mathbf{X}(i')$ to be the set of truck types which are compatible with loader type i'. Next we define a constraint that ensures that all equipment compatible with a particular loader type can satisfy the requirements:

$$\sum_{i \in \mathbf{X}(i'),l} P_i^{k,l} \mathbf{x}_i^{k,l} \geq \sum_{l} P_{i'}^{k,l} \mathbf{x}_{i'}^{k,l} \qquad \forall \quad i' \in \mathbf{X}', k. \tag{6.10}$$

However, we must also consider the possibility that two loader types can be selected. Equation (6.10) only accounts for the case where all trucks are compatible with one loader, i'. If we just have constraints (6.10), then this could lead to double counting. We want to consider the loader type pairs case (i', h') and ensure that the compatible fleet of trucks can service both of these loaders together. We denote the union of the compatible truck fleets by the set $\mathbf{X}(i', h')$:

$$\sum_{i \in \mathbf{X}(i',h'),j,l} P_i^{k,l} \mathbf{x}_{i,j}^{k,l} \geq \sum_{l} (P_{i'}^{k,l} \mathbf{x}_{i',j}^{k,l} + P_{h'}^{k,l} \mathbf{x}_{h',j}^{k,l}) \qquad \forall \quad (i', h') \in \mathbf{X}', k. \tag{6.11}$$

Similarly, we must allow the possibility of three types of loaders, (i', h', j'):

$$\sum_{i \in \mathbf{X}(i',h',j'),j,l} P_i^{k,l} \mathbf{x}_{i,j}^{k,l} \geq \sum_l (P_{i'}^{k,l} \mathbf{x}_{i',j}^{k,l} + P_{h'}^{k,l} \mathbf{x}_{h',j}^{k,l} + P_{j'}^{k,l} \mathbf{x}_{j',j}^{k,l}) \qquad \forall \ (i', h', j') \in \mathbf{X}', k. \tag{6.12}$$

Note that the assumption of full period utilisation may cause some problems with these constraints. In particular, we could be forcing the trucking fleet to exceed the productivity requirements of the mine—if the loaders exceed the productivity requirements we don't want to force the trucks to match them. We can rectify this by introducing an indicator variable, $\mathbf{h}_{A'}^k$, where $A' \subset \mathbf{X}'$, that will choose one of the following two constraints to dominate:

$$\sum_{i \in \mathbf{X}(A'),l} P_i^{k,l} \mathbf{x}_i^{k,l} \geq \sum_{i' \in A',l} P_{i'}^{k,l} \mathbf{x}_{i'}^{k,l} - M\mathbf{h}_{A'}^k \qquad \forall \ A' \subset \mathbf{X}', k, \tag{6.13}$$

$$\sum_{i \in \mathbf{X}(A'),l} P_i^{k,l} \mathbf{x}_i^{k,l} \geq T^k \mathbf{h}_{A'}^k \qquad \forall \ A' \subset \mathbf{X}', k. \tag{6.14}$$

Because we are taking the power set of $\mathbf{X}'$, this will generate $2K(2^N - 1)$ constraints where K is the total number of periods and N is the number of loaders). For the 13 period, 27 loader case study this equates to 35 billion constraints. However, the number of loaders selected in the final solution will generally be much lower than the complete set. Thus we can limit the generation of constraints by allowing a maximum of α loader types. This will produce $2K\binom{h}{\alpha}$ constraints. For the 13 period, 27 loader problem with a maximum of 4 loader types, this equates to just 542,178 constraints.

We can further reduce the number of constraints in the model by entering this constraint set using a separation algorithm. That is, these constraints are only entered into the model if they are violated by the solution of the model. This is a useful technique for eliminating a large set of inactive constraints from the model: potentially improving the computation time and solvability of the model, and freeing up a large quantity of memory that would have been required to store the model in the computer.

6.2.4.4 Variable Transition

Much of the linearisation in this model is due to the way that we have defined the decision variables: we capture the time period and age (in utilised age brackets) of each piece of equipment in respective indices, k and l. It is important to establish the relationship between these two indices (Fig. 6.4).

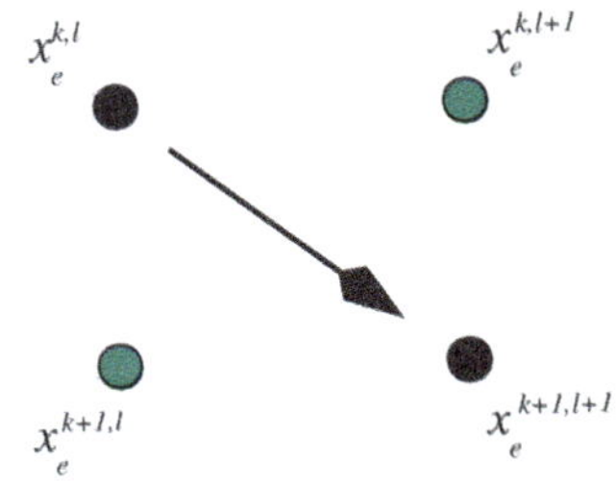

Fig. 6.4 The relationship between the indices k and l for the variable transition constraint. Under the assumption of full utilisation, an increment in time period (k) results in an increment in age (l)

We assume that equipment is fully utilised, so all equipment gains a value of 1 on each index k each period. However, equipment can be salvaged at the start of every period. We express this relationship for all equipment types in constraint (6.15):

$$\mathbf{x}_e^{k,l} = \mathbf{x}_e^{k-1,l-1} - \mathbf{s}_e^{k,l} \qquad \forall \quad k > 0, l \in [1, L^k(e)], e. \tag{6.15}$$

6.2.4.5 Forced Salvage

The maximum age of each type of equipment can vary substantially: from 25,000 to 100,000 h. When a piece of equipment reaches its maximum age, we force it to retire at the beginning of the following period.

$$\mathbf{x}_e^{k,l} = \mathbf{s}_e^{k+1,l+1} \qquad \forall \quad l > L^k(e), e, k. \tag{6.16}$$

During variable creation we can prevent over-age variables from existing in the first place, thus effecting forced salvage.

6.2.4.6 Salvage Restriction

In this formulation we profit from the salvage of trucks and loaders. We must therefore prevent the salvage of equipment that is not owned:

$$\mathbf{x}_e^{k-1,l-1} \geq \mathbf{s}_e^{k,l} \qquad \forall \quad k > 0, l \in [1, L^k(e) + 1], e. \tag{6.17}$$

We also set unbounded salvage variables to 0 to prevent them from dominating the objective function:

$$\mathbf{s}_e^{k,0} = 0 \qquad \forall \quad e, k. \tag{6.18}$$

6.2.4.7 Pre-existing Equipment

A novelty of this model is the ability to include pre-existing equipment in the optimisation process. If we have $\mathbf{x}_e^{P(e)}$ equipment of type e which are of age $P(e)$, then we either use or salvage these equipment.

$$\mathbf{x}_e^{0,P(e)} + \mathbf{s}_e^{0,P(e)} = x_e^{P(e)} \qquad \forall \quad e \in \mathbf{P}. \tag{6.19}$$

Pre-existing equipment that has exceeded the maximum age, $L(e)$, for its equipment type, e, must be salvaged immediately:

$$\mathbf{s}_e^{0,P(e)} = x_e^{P(e)} \qquad \forall \quad e \in \mathbf{P},\ P(e) > L^k(e). \tag{6.20}$$

6.2.5 Complete Model

$$\min \quad \sum_{e,k} F_e D_1^k \mathbf{x}_e^{k,0} + \sum_{e,k,l,h} B_{h,e}^{k,l} D_1^k V_e^{k,b+h-1} \mathbf{x}_e^{k,l} - \sum_{e,k,l} F_e D_2^{k,l} \mathbf{s}_e^{k,l} \tag{6.21}$$

$$\text{s.t.} \quad \sum_{i',l} P_{i'}^{k,l} \mathbf{x}_{i'}^{k,l} \geq T^k \qquad \forall \quad i' \in \mathbf{X}', k, \tag{6.22}$$

$$\sum_{i \in \mathbf{X}(A'),l} P_i^{k,l} \mathbf{x}_i^{k,l} \geq \sum_{i' \in A',l} P_{i'}^{k,l} \mathbf{x}_{i'}^{k,l} - M \mathbf{h}_{A'}^k \qquad \forall \quad A' \subset \mathbf{X}', k, \tag{6.23}$$

$$\sum_{i \in \mathbf{X}(A'),l} P_i^{k,l} \mathbf{x}_i^{k,l} \geq T^k \mathbf{h}_{A'}^k \qquad \forall \quad A' \subset \mathbf{X}', k, \tag{6.24}$$

$$\mathbf{x}_e^{k,l} = \mathbf{x}_e^{k-1,l-1} - \mathbf{s}_e^{k,l} \qquad \forall \quad k > 0, l \in [1, L^k(e)], e, \tag{6.25}$$

$$\mathbf{x}_e^{k,l} = \mathbf{s}_e^{k+1,l+1} \qquad \forall \quad l > L(e), e, k, \tag{6.26}$$

$$\mathbf{x}_e^{k-1,l-1} \geq \mathbf{s}_e^{k,l} \qquad \forall \quad k > 0, l \in [1, L^k(e) + 1], e, \tag{6.27}$$

$$\mathbf{s}_e^{k,0} = 0 \qquad \forall \quad e, k, \tag{6.28}$$

$$\mathbf{x}_e^{0,P(e)} + \mathbf{s}_e^{0,P(e)} = x_e^{P(e)} \qquad \forall \quad e \in \mathbf{P}, \tag{6.29}$$

$$\mathbf{s}_e^{0,l} = x_e^{P(e)} \qquad \forall \quad e \in \mathbf{P}, P(e) > L(e), \tag{6.30}$$

$$\mathbf{x}, \mathbf{s} \in \mathbb{Z}^+,$$

$$\mathbf{h} \in \{0, 1\}.$$

6.3 Computational Study

For our experiments, we consider the *many locations* case study presented in Chap. 5, and aggregate the production requirements to obtain a single location. We implemented the case study for 13 periods with 276612 variables and 7910 constraints. We programmed the model in *C++* using *Ilog Concert Technology v2.5* objects, and solved the program with default IP algorithms in *Ilog Cplex v11.0*. We performed these tests on a Pentium 4 PC with a 3.0 GHz CPU and 2.5 GB of RAM.

We present the summary of results in Table 6.1.

Table 6.1 Summary of results for the 13-period case study with varying depreciation

Depreciation (%)	Time (s)	Objective function ($)
40	3255	1.25806×10^8
50	9873	1.26075×10^8
60	3336	1.26212×10^8

The optimal cost of 1.25806×10^8 was obtained after 3255 s (2.5 h) for the 40% depreciation problem. Convergence for this problem was swift, reaching within 1% of optimality in just 15 s. Figure 6.5 depicts the rate at which the *best integer* solution converges with the *best node* solution.

In our solution (Fig. 6.6), two 60-ton loaders were selected over the course of the schedule. Five different types of trucks were selected to work with these loaders: a 136-ton truck, three 177-ton trucks, two large 230-ton trucks, twenty 150-ton trucks and 11 pre-existing trucks. We calculated the *match factor* for each fleet (adopting the heterogeneous formula in Chap. 4). The values we calculated for each fleet exceeds 1 in each period. This demonstrates that the optimal solution is to work the loaders as much as possible and permit the trucks to wait in a queue. This solution is a reflection of the large cost of operating high capacity loaders compared to the cost of operating haul trucks.

Our industry partner derived three solutions for this case study which we now use for a retrospective comparison. The first is the solution provided by an inhouse equipment selection spreadsheet tool. The loader solution kept the youngest pre-existing loader with capacity 34-tons, and purchased two new 40-ton loaders. The truck solution is presented in Table 6.2. This solution cost 1.51483×10^8. Our integer programming model improved this by 17.7%—an increase in profit of $26.9 million.

A second solution was produced by an equipment selection manager (for the industry partner), who recommended the salvage of all pre-existing loaders. One

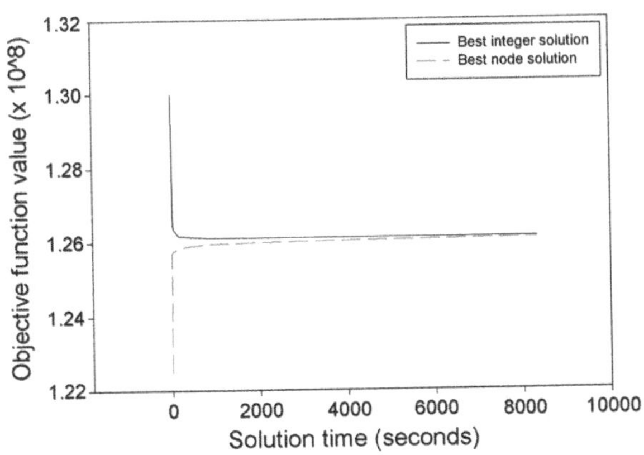

Fig. 6.5 Convergence of the 13-period case study for the single-location model

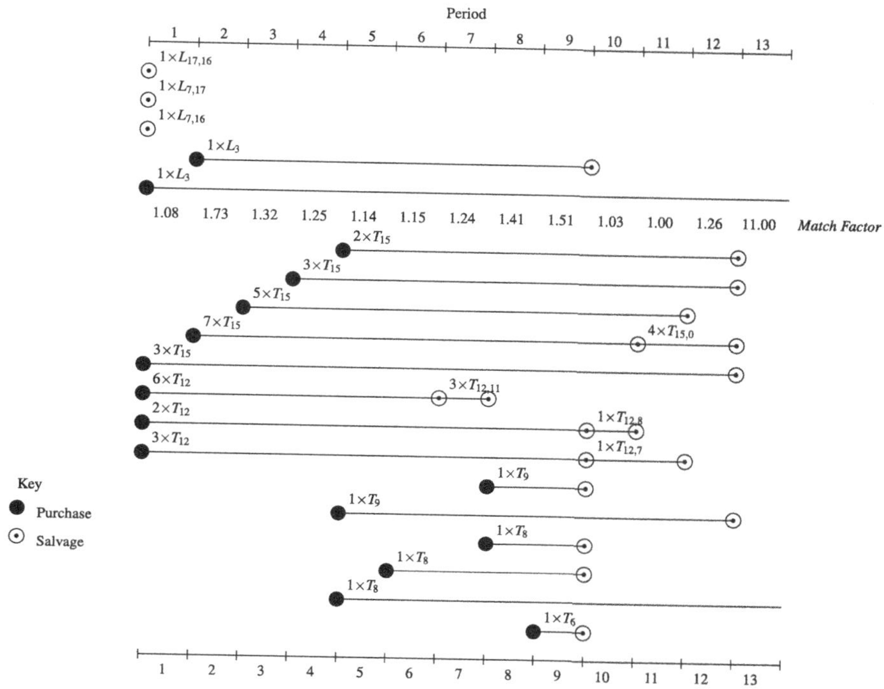

Fig. 6.6 The optimal solution for the case study

Table 6.2 The retrospective truck purchase and salvage policy

	Periods												
	1	2	3	4	5	6	7	8	9	10	11	12	13
Pre-existing T_{12}	11	11	11	11	11	11	11	11	11	11	11	11	11
New T_{12}	5	5	5	5	5	5	5	5	5	5	5	5	5
New T_{12}	0	2	2	2	2	2	2	2	2	2	2	2	2
New T_{12}	0	0	0	0	1	1	1	1	1	1	1	1	1

42-ton loader and two 57-ton loaders were to be purchased. The same truck solution as presented in Table 6.2 was to be used. This solution cost $\$1.55241 \times 10^8$. Our model solution provides an improvement of 18.75%, or \$29.1 million.

A third solution was provided, which was the actual solution adopted at the time. All pre-existing loaders were salvaged and three 57-ton loaders were purchased. The truck purchase and salvage policy from Table 6.2 was used. This solution cost $\$1.66550 \times 10^8$. Our model improved this solution by 24.3%, amounting to a \$40.4 million cost difference.

Clearly, our solution (although more complicated) is by far the cheaper solution, indicating the advantages of applying an integer programming model to this problem.

However, we cannot claim that our solution is definitely superior as we did not consider the cost of auxiliary equipment. In particular, our solution requires the operation of multiple types of equipment which may increase the need for auxiliary equipment and other costs. The policy that was actually used avoids costs associated with multiple equipment types by adopting homogeneous fleets.

6.4 Conclusion

In this chapter, we have presented a computationally fast integer program that can outperform industry generated solutions substantially. We have placed particular importance on pre-existing equipment and compatibility in this model. Ensuring compatibility of multiple types of equipment is a unique aspect of this model that allows greater freedom for the equipment selection manager for two reasons:

- the manager may consider purchasing equipment that is different to any of the pre-existing fleet;
- the manager may consider purchasing mixed fleets that better suit the productivity requirements of the mine, and consequently may achieve lower operating expenses.

The ability to include pre-existing equipment in the equipment selection process is novel for the mining industry.

Another important advantage of this model is that it appropriates the operating cost across brackets. That is, within one period, if a piece of equipment graduates to the next age bracket (relative to its age), then the operating cost will reflect this in a proportional manner.

The solutions presented in this chapter test the "bigger is better" philosophy that is commonplace in the mining industry. In our study, bigger equipment is considered too costly to select all the time. Instead the solutions tend toward moderate sized equipment that can adapt to changing productivity requirements and truck cycle times. This result suggests that bigger machines provide a cost benefit only if they can be fully utilised for the entire period.

Although the assumptions contribute to the tractability of the model, they can also detract from its realism. In particular, the assumptions of *full period utilisation* and *single mining location* are abstractions of the problem, designed to create a simple model.

The validity of the assumption of full period utilisation is wholly dependant on the management practices of the mine. Some mines may only operate the equipment as it is needed, while others will operate all equipment that is owned but at lesser capacity. The full period utilisation assumes the latter is taking place. This can lead to peculiar solutions, such as having many different truck types (up to 5 in some cases), with some equipment salvaged after only one year of ownership. This can be attributed to the cost of operating any owned equipment for the entire period. This forces the selection of truck types that closely fit the productivity requirements, and with the presented data set this sometimes results in many truck types being selected.

However, hiring equipment may offset this problem, and encourage the model to select a more manageable variety of truck types. We derive a utilisation-based model in Chap. 8.

The productivity constraints assume that no bunching occurs and the selected fleets operate efficiently (though not necessarily to full capacity). This is not an accurate abstraction of reality. For example, if we had 20 trucks operating in a fleet with one loader operating to full capacity and we added 5 more trucks, then the truck cycle time would increase due to bunching. Thus, with our present model of the problem, the productivity of the system will be inaccurately overestimated in this case. In actuality, the bunching is a function of the fleet solution. It would be a great challenge to pursue an integer programming model that can capture this type of cyclic relationship.

The assumption of *single mining location* is restrictive in the sense that this model cannot be applied to a mine with an intricate system of loading locations and trucking routes with guaranteed satisfaction of productivity requirements. This model is therefore useful for mining schedules that are not fully developed, or for mines that do not have much movement between alternate dump-sites. An obvious extension is to relax the assumption by allowing multiple loading sites, dumping sites and truck routesi. We provide this extension in Chap. 7.

Notation Index

A summary of the notation used in this chapter is as follows.

- $\mathbf{X}$ the set of all available truck types.
- $\mathbf{X}'$ the set of all available loader types.
- i the truck type index, $i \in \mathbf{X}$.
- i' the loader type index, $i' \in \mathbf{X}'$.
- e the equipment type index, $e \in \mathbf{X} \cup \mathbf{X}'$.
- j the location index.
- k the period index.
- l the equipment age index.
- $\mathbf{P}$ the set of all pre-existing equipment types.
- $\mathbf{X}(A')$ the set of all truck types compatible with loaders $A' \in \mathbf{X}'$.
- K the total number of periods in the mine plan.
- J the total number of locations and/or routes in the mine plan.
- L the maximum age (in operating hours) of all equipment.
- $L(e)$ the maximum age (in operating hours) unique for each truck and loader type, e.
- $\mathbf{x}_e^{k,l}$ the number of equipment type e selected in period k, aged l.
- $\mathbf{f}_{e,j}^{k,l}$ the proportion of equipment type e selected in period k, aged l sent to work on route j.
- $\mathbf{s}_e^{k,l}$ the number of equipment type e salvaged in period k, aged l.
- $\mathbf{h}_{A',j}^{k}$ the indicator variable for constraint dominance for loader[s] type $A' \in \mathbf{A}$ in period k for location j.

$x_e^{P(e)}$ the number of pre-existing equipment type e aged $P(e)$ at the start of the schedule.
F_e the fixed cost (capital expense) of obtaining equipment type e.
D_1^k the discount factor in period k.
$D_2^{k,l}$ the discount factor in period k and depreciation factor for equipment aged l.
I the fixed interest rate used in the net present value discount factor.
F_e the fixed cost of purchasing equipment of type e.
B_0 the size of age brackets.
$a_e^{k,l}$ the availability of equipment type e, aged l in period k.
$B_{h,e}^l$ the proportion of time equipment type e aged l spends in cost bracket $h \in \{1, 2\}$. i
$V_e^{k,l}$ the variable expense for equipment type e, aged l in period k.
$P_e^{k,l}$ the productivity of equipment e, in period k at age l.
T_j^k the required productivity of the mine (in tonnes) for period k at location j.
$P(e)$ the highest starting age of any pre-existing equipment of type e.
M a large number.

Acknowledgements Components of this chapter were presented in [1] and [2].

References

1. C. Burt, An optimisation approach to materials handling in surface mines. Ph.D. thesis, Mathematics and Statistics, Curtin University of Technology, Bentley, Perth, Australia, 2008. http://espace.library.curtin.edu.au:80/R?func=dbin-jump-full&local_base=gen01-era02&object_id=165737
2. C. Burt, L. Caccetta, P. Welgama, L. Fouché, Equipment selection with heterogeneous fleets for multiple period schedules. J. Oper. Res. Soc. **62**, 1498–1509 (2011)

Chapter 7
Multiple Locations Equipment Selection

Christina N. Burt, Louis Caccetta, Leon Fouché and Palitha Welgama

7.1 Introduction

In this chapter, we consider the equipment selection problem in the context of a multiple location mine. In particular, we wish to select a compatible fleet of trucks and loaders to move mined materials between multiple mining and dumping sites, at minimum cost. As far as we know, there has been no literature (other than our work) that addresses multiple period schedules, multiple locations and allows for pre-existing equipment (and subsequent heterogeneous fleets). Multiple periods and locations increase the dimension of the problem, but it is the pre-existing equipment that requires compatible fleet constraints that makes the problem difficult. We seek to address these deficiencies in the literature. The main features of our model approach are:

- the consideration of a multiple location and multiple period mining plan;

The original version of this chapter was revised: Originator type has been changed and the chapter author names have been included. The erratum to this chapter is available at https://doi.org/10.1007/978-3-319-76255-5_11

C. N. Burt (✉)
Department of Mathematics and Statistics, The University of Melbourne, Parkville, VIC, Australia
e-mail: c.burt@ms.unimelb.edu.au

L. Cacceta
Department of Mathematics and Statistics, Curtin University, Bentley, WA, Australia
e-mail: L.Caccetta@curtin.edu.au

L. Fouché · P. Welgama
Rio Tinto Iron Ore, Perth, WA, Australia
e-mail: Leon.Fouche@riotinto.com

P. Welgama
e-mail: palitha.Welgama@riotinto.com

C. N. Burt and L. Caccetta (eds.), *Equipment Selection for Mining: With Case Studies*, Studies in Systems, Decision and Control 150,
https://doi.org/10.1007/978-3-319-76255-5_7

- the consideration of heterogeneous fleets and subsequent compatibility requirements;
- simultaneously optimising the purchase and salvage policy, and equipment scheduling policy (i.e. allocation policy);
- providing correction for discretisation error;
- variable preprocessing based on a 'staircase' structure in the solution;
- development of solving approaches that help to reduce the total number of constraints in the model;
- illustration of computation effectiveness in a real-world context through two case studies.

In the remainder of this section, we will outline some important background to the multiple location equipment selection problem and discuss relevant related literature.

A typical surface mine may have several mining locations, several dump-sites or several routes from a location to a dump-site, as illustrated in Fig. 7.1. Different mining locations may have capacity requirements that affect the type of loader selected to excavate the material, and consequently alter the rate of production at that site. Along the routes the trucks may alternate between the mill and dump sites; the routes themselves may also vary significantly in terms of time required to perform a full cycle. Furthermore, the consideration of multiple locations or routes introduces the need to allocate equipment to locations. Since the movement of equipment around a network must be consistent, this feature leads to an underlying problem which can be described as a multi-commodity flow problem [2]. Purchase may occur in any period. The problem is therefore at least as difficult as the fixed-charge, capacitated multi-commodity flow problem. To see this, consider Fig. 7.2. In a multi-commodity flow problem defined on a network, there may be several sources, here s, and several sinks, here t. If we consider trucks moving along routes between sources and sinks in a single commodity context, then this would be equivalent to fixing homogeneous equipment to routes. However, in a multi-commodity flow problem, there may also be more than one type of *commodity*, or flow, leaving from any source. Furthermore,

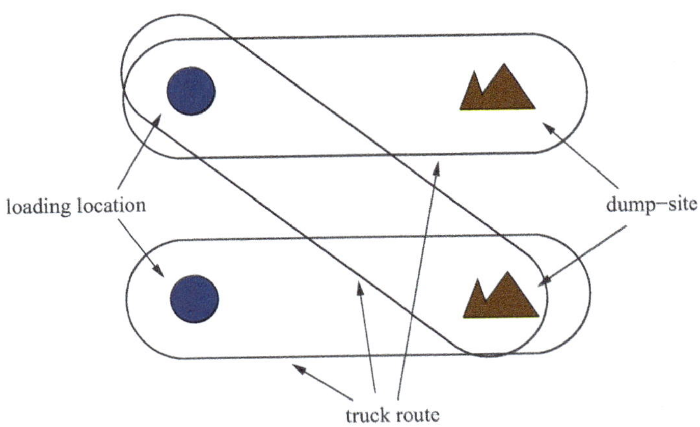

Fig. 7.1 A multiple-location mine model with 2 loading locations, 2 dump-sites and 3 truck routes

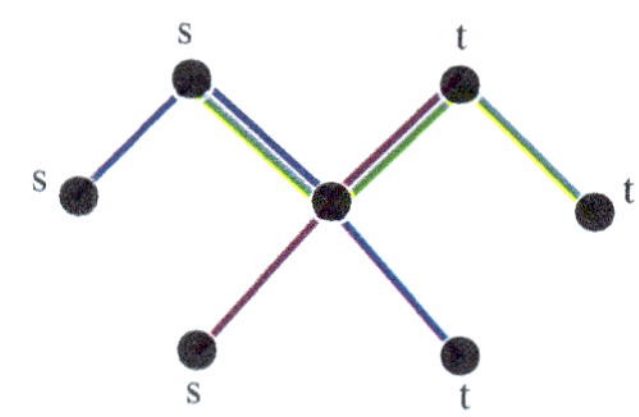

Fig. 7.2 A simple multi-commodity flow network

the flow can split at a node. Carrying the analogue to equipment selection, this is equivalent to allocating heterogeneous truck flow on routes, where the flow may separate and switch to other sinks or sources. We obtain *fixed-charge* and *capacitated* from charging a fixed price for equipment at the time of purchase, and limiting the number of equipment we may use in flow by fleetsize.

The first mine plan must use approximate transportation costs, since the equipment selection solution will not be known. Over the life of the mine, the generation of mine plans at planning intervals will depend on the selected equipment fleet. Therefore, new plans should be generated if changes in the equipment fleet are generated at an equipment planning interval.

In the surface mining industry, the generally non-linear non-convex operating cost and availability (as functions of the age of the equipment) are commonly discretised to step-wise functions that are divided into age brackets of size B_0, as illustrated in Fig. 7.3. The operating expense reflects the cost of operating and maintaining the equipment. It takes into account varying maintenance expenses, availability and productivity levels, which are known to vary with the age of the equipment. In our case studies, we use an age bracket size of 5000 h (as in [4]). We have similar factors for the availability of the equipment (the proportion of time it is available to work), utilisation (the proportion of time it is effective) and maintenance (the proportion of time the equipment is available after maintenance). All of these factors

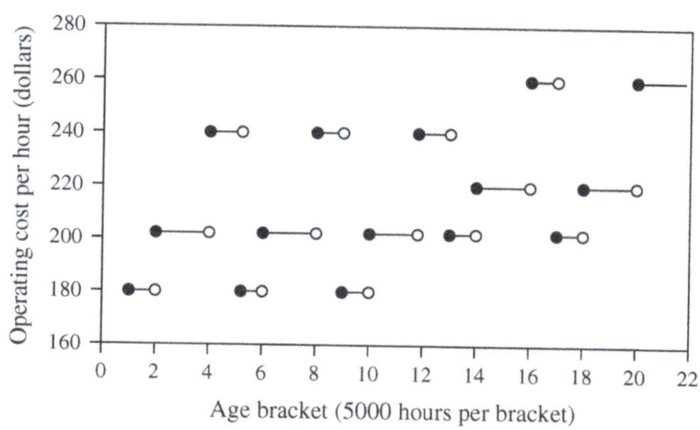

Fig. 7.3 Discretised operating cost function against age brackets. The rise in operating cost reflects the increased maintenance expense; large drops in the expense occur when a significant maintenance, such as overhaul, has taken place

are functions of the age bracket the equipment is in, indicating that the performance of the equipment changes with its use in a non-linear way.

The full period utilisation assumption is really about the granularity of our data—we have set up this problem with time windows of annual periods, and there are considerations such as budgets and labour flexibility which are typically annual in nature. Purchase and salvage of equipment typically occur periodically, so accordingly we define a period to be 1 year in length as a reasonable timeframe with which to consider purchase and salvage of equipment and any cost benefits associated with bulk purchases. Smaller periods may be considered, such as quarterly, to generate finer fidelity schedules and allocation solutions, but purchase and salvage would generally not occur at these smaller intervals. For this reason, and also to aid the tractability of the problem, we will not consider periods smaller than 1 year in length in our experiments.

Often the data for costing and even projected demand is imperfect. Since the projection period itself can be quite long (over 20 years), a pragmatic approach is to formulate the problem as a deterministic problem to match the provided mine plan, and perform sensitivity analysis to understand the robustness of the obtained solutions. Furthermore, since the problem is already large-scale, the additional consideration of stochasticity and uncertainty would only exacerbate the difficulty of the problem. In this sense, obtaining solutions via a deterministic modeling approach such as mixed-integer programming is appropriate for the multi-location equipment selection problem where truck and loader units are integral and the capacity constraints can be captured linearly.

We derive our model in Sect. 7.2. We begin by describing the model setting, decision variables and reductions, before deriving the objective function with corrector for discretisation error (Sect. 7.2.3) and constraints (Sect. 7.2.4). We describe two surface mining case studies with varying mining locations and routes in Sect. 7.3 with a note on practical implications of our solutions (Sect. 7.3.3). We conclude with a discussion of the research and possible further advancements in Sect. 7.4.

7.2 The Model

7.2.1 *Assumptions*

We restrict the focus of this chapter to mining equipment selection under the following assumptions:

- Multiple locations—Multiple locations and multiple routes exist on which the selected fleet may move about;
- Multiple truck cycle times—Truck cycle time is fixed for a given route, where the route is defined as a pair of loading location and dumping destination;
- Known mine plan—An acceptable mine plan has already been derived (including selection of mining method), and is fixed for optimisation period;

- Salvage—All equipment is salvageable at the start of each period at some depreciated value of the original capital expense;
- No auxiliary equipment—Wheel loaders and small trucks are not considered in this model, although can be easily included if the cost and maintenance data is available;
- Known operating hours—The operating hours of the mine are estimated by taking planned downtime, blasting and weather delays into account;
- Heterogeneous fleets—Different types of equipment may work within one fleet, so long as compatibility requirements are satisfied;
- Fleet retention—All equipment is retained at the end of the last period;
- Full period utilisation—Operating costs are charged as though the equipment has been fully utilised for each entire period in which it is owned.
- Age bracket size—The size of an age bracket, B_0, used to discretise the availability function, is strictly larger than the size of a period, i.e. $B_0 > \max\{H^k\}$;
- Equipment availability—Equipment availability, maintenance requirements and equipment utilisation change over time, and can be effectively approximated using age brackets.

7.2.2 Decision Variables and Notation

We provide a summary of notation at the end of this chapter. We define the variables for our model using an arc-based representation of the mining locations ($i \in I$) and routes, (i, j), to dumpsites. To simplify notation, we denote a route by $j \in J$. We denote the set of all truck types by T and loader types by L. Here we adopt three indexes to represent type ($t \in T$), period ($k \in \{1, 2, \ldots, K\}$) and age bracket ($m \in \{1, 2, \ldots, M\}$). We use integer variables to track whole equipment units while continuous variables allocate equipment to routes:

$\mathbf{x}_{t,k,m}$: number of trucks of type t owned in period k which are in age bracket m (integer variable);

$\mathbf{y}_{l,k,m}$: number of loaders of type l owned in period k which are in age bracket m (integer variable);

$\mathbf{s}_{t,k,m}$: number of trucks of type t salvaged in period k which are in age bracket m (integer variable);

$\mathbf{s}_{l,k,m}$: number of loaders of type l salvaged in period k which are in age bracket m (integer variable);

$\mathbf{f}_{t,j,k,m}$: portion of trucks of type t, in age bracket m, that are allocated to route j in period k, where $\mathbf{f}_{t,j,k,m} \in [0, \mathbf{x}_{t,k,m}]$ (continuous variable),

$\mathbf{f}_{l,i,k,m}$: portion of loaders of type l, in age bracket m, that are allocated to location i in period k, where $\mathbf{f}_{l,i,k,m} \in [0, \mathbf{y}_{l,k,m}]$ (continuous variable).

To simplify prose in this chapter, we sometimes just describe constraints for the trucks if a corresponding, and identical, constraint also exists for the loaders in our model. The complete model (Sect. 7.2.5) contains all constraints. The relationship

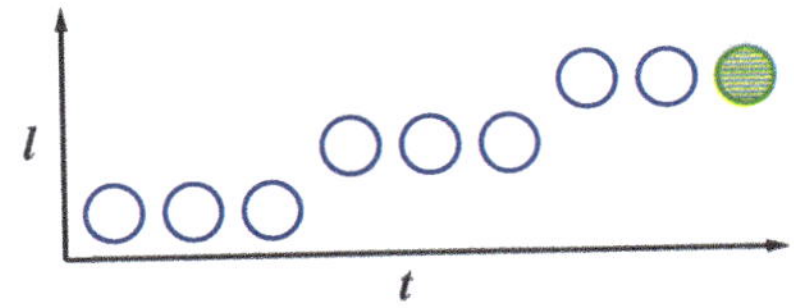

Fig. 7.4 As time progresses, the $\mathbf{x}_{t,k,m}$ variables move along the 'steps' of age-brackets, ending with a single instance of the variable $\mathbf{s}_{t,k,m}$. This staircase structure is important for preprocessing

between $\mathbf{x}_{t,k,m}$ and $\mathbf{s}_{t,k,m}$ is illustrated in Fig. 7.4. The $\mathbf{f}_{t,j,k,m}$ variable will trace the 'staircase' and allocate portions of the total time to particular routes, as described in the following section. The analogue exists for the loader variables, though for brevity we restrict the description to trucks.

The precedence characteristic of the solutions elicit a natural 'staircase' structure in the possible values of the decision variables. The possible height and depth of the staircase is limited by the maximum number of hours the equipment can be used per period in combination with the age of the pre-existing equipment. We use $M_t^{\max}$ to denote the maximum age bracket of truck type t—this value may vary depending on the equipment type. Since $B_0 > H_k$ (the age-bracket size is greater than the size of the period), $M_t^{\max}$ can also be restricted by the time period as equipment cannot age more than one age bracket in one period. However, we consider the possibility of pre-existing equipment (which is known *a priori*), with the starting age $M_t^{\max}$. The maximum age bracket, $M_k(t)$, for truck type t in time period k is as follows:

$$M_k(t) = \min\{M_t^{\max} + k - 1, M_t^{\max}\}.$$

This becomes a reduced limit for index l in any relevant constraint. We note that as salvage occurs at the start of the period, salvage variables extend to $M_k(t) + 1$.

At implementation, we do not permit over-age variables from existing, thus effecting forced salvage. That is, restricting variable creation in this way is equivalent to the following constraint:

$$\mathbf{x}_{t,k,m} = \mathbf{s}_{t,k+1,m+1} \qquad \forall\ m > M_k(t), t \in T, k \in \{1, \ldots, K-1\}.$$

If we consider the set of pre-existing equipment types, $\mathbf{P}$, there is the possibility of immediate salvage in the age bracket $m = 1$. However, for all other equipment types we can prevent the unbounded salvage variables from dominating:

$$\mathbf{s}_{t,k,1} = 0 \qquad \forall\ t \notin \mathbf{P}, k \in \{1, \ldots, K\}.$$

This too can be effected during variable creation, thereby reducing the overall number of variables and constraints required to represent the problem.

7.2.3 Objective Function

We must satisfy demand either by location or route, depending on the nature of the mine plan. For this formulation, the availability of the equipment throughout the period is determined by the age bracket. The *production capability* is determined by its availability ($A_{t,m}$), capacity (C_t) and cycle time ($\tau_{t,j}$) (where the cycle time for a truck is the route cycle time and the cycle time for a loader is the time required to fill a particular truck type):

$$P_{t,j,k,m} = \frac{A_{t,m} C_t}{\tau_{t,j}}. \quad (7.1)$$

In our objective for the mixed-integer program, we minimise the cost of running the fleet for the entire mine plan, including purchase, operating expense and salvage. We represent the fixed cost of purchasing truck of type t by F_t and discount this purchase to the present using a discount factor,

$$D_k^1 = \frac{1}{(1+I)^k}$$

(where k is the current period, starting from 1, and I is the interest rate). Thus, the *total capital expense* for a truck of type t is

$$\sum_{t,k} F_t D_k^1 \mathbf{x}_{t,k,1},$$

with a corresponding term for loaders.

The discretisation of the variable costs over age brackets can lead to misleading operating costs, as the equipment may start the period in one age bracket (and corresponding cost) and move into another age bracket for the remainder of the period. To illustrate this, consider Fig. 7.5. The best case is case (a), where the age bracket of the equipment is correct for the entire period. However, in case (b) the equipment moves into a new age bracket during the period. This will result in a discretisation error in the model.

In order to provide the most accurate costing possible for this model, we must determine the proportion of time that the equipment remains in age bracket m within

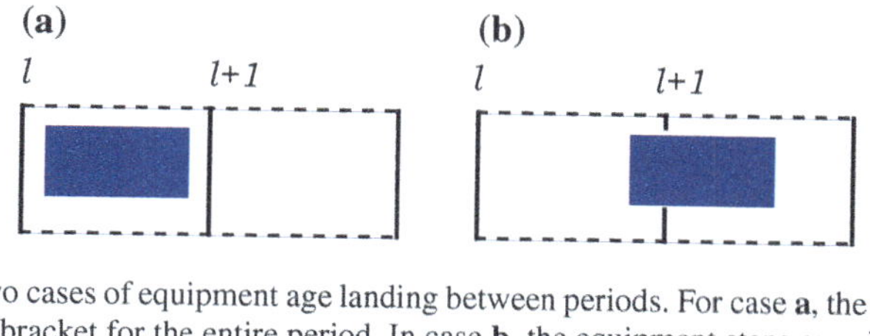

Fig. 7.5 The two cases of equipment age landing between periods. For case **a**, the equipment stays in the same age bracket for the entire period. In case **b**, the equipment steps over into the next age bracket within the period

the period, and the proportion it lies in the proceeding age bracket, $m + 1$. The process to achieve this in a way that obtains a constant coefficient for variables, thereby maintaining linearity in the model, is described over the next page. Although it appears tedious and cumbersome, in practice it is very simple to implement in a computer program.

To calculate the portion of time that each set of equipment spends in each age bracket, we first calculate the age of the equipment (in age brackets) in any given period. To do this, we need to know when the equipment was purchased. We denote the purchase period by k'. The equipment must also be owned in the current period for this calculation to take place. The availability of equipment is the proportion of the period that the equipment is available to operate—unavailability is often due to planned maintenance. Availability as it is used here is calculated using the availability, utilisation and maintenance factors discussed above, and is dependent on the current age bracket of the equipment. Let H_k be the operating hours for period k and $A_{t,m}$ be the availability of truck type t in age bracket m. We obtain the age of equipment in operated hours using a recursive formula since the availability of the equipment is a function of equipment age itself. The base of the recursion is:

$$\beta(k') = A_{t,1} H_{k'}.$$

Then, the age of the equipment in operated hours can be obtained for any period k by:

$$\beta(k) = \sum_{k' \leq h' < k} A_{t,\left\lfloor \frac{\beta(h'-1)}{B_0} \right\rfloor} H_{h'}.$$

We obtain the age bracket in which equipment lies at the beginning of a period by $b(k)$:

$$b(k) = \left\lfloor \frac{\beta(k)}{B_0} \right\rfloor.$$

Since $B_0 > H_k$, the equipment may only lie in $h \in \{1, 2\}$ age brackets within one period. To begin, we define $B^h_{t,k,m}$ to be the proportion of total operated hours that truck t spends in age bracket $m + h - 1$ in period k (where the incumbent age bracket is m).

Theorem 7.1 *The proportion of time that any group of machinery spends in any one age bracket can be represented by the following two expressions:*

$$B^1_{t,k,m} = \begin{cases} 1, & \text{if } (m+1)B_0 - \beta(k) > A_{t,b(k)} H_k \\ \dfrac{(m+1)B_0 - \beta(k)}{A_{t,b(k)} H_k}, & \text{otherwise} \end{cases}$$

and

$$B^2_{t,k,m} = 1 - B^1_{t,k,m}.$$

Proof The age at the start of the period is given by $\beta(t)$ as defined above. The quantity of hours worked in the current period is given by:

$$A_{t,m} H_k.$$

If the equipment stays in the same age bracket for the entire period, we require that the difference between the marker for the next age bracket, $(m+1)B_0$, and the age at the start of the period exceeds the quantity of hours worked in the current period. That is, if:

$$(m+1)B_0 - \beta(k) > A_{t,b(k)} H_k.$$

Similarly, if the equipment moves into another age bracket for part of the period, we can simply look at the difference between the marker for the next age bracket and the age at the start of the period. Dividing by the operated hours for the current period gives the proportion of total operated hours, as required. ■

We can easily adjust these formulas for the case of pre-existing equipment, but for the sake of clarity omit this here. We can now use $B^1_{t,k,m}$ and $B^2_{t,k,m}$ to correct the operating cost (denoted by $V_{t,k,m}$) in the objective function as follows (also with a discount factor):

$$\sum_{t,k,m,h} \frac{B^h_{t,k,m}}{(1+m)^k} V_{t,k,b(k)+h-1} \mathbf{x}_{t,k,m}.$$

Lastly, we consider the income from salvaging old equipment. We apply a combined depreciation (at rate J per period) and discount factor (at rate I per period):

$$D^2_{k,m} = \frac{(1-J)^l}{(1+I)^k},$$

where l is the age of the equipment at the start of period t. Since F_e is the original capital expense, the salvage 'cost' is:

$$-\sum_{t,k,m} F_t D^2_{k,m} \mathbf{s}_{t,k,m},$$

with a corresponding term for loaders.

7.2.4 Constraints

We require the loaders to satisfy the production demand, $D_{i,k}$, at location $i \in I$, giving us the following *capacity demand constraints*:

$$\sum_{l,m} P_{l,k,m}\mathbf{f}_{l,i,k,m} \geq D_{i,k} \quad \forall\ k \in \{1, \ldots, K\}, i \in I, \tag{7.2}$$

where $P_{l,k,m}$ is the maximum possible productivity of loader l that is aged m in period k. This expression is obtained by considering equipment capacity, swing time (time to deliver one load to the truck), number of required swings (truck to loader capacity ratio), and downtime due to maintenance and other factors. Similarly, we require the trucks to satisfy the demand for each route or dump site, $j \in J$:

$$\sum_{t,m} P_{t,k,m}\mathbf{f}_{t,j,k,m} \geq D_{j,k} \quad \forall\ k \in \{1, \ldots, K\}, j \in J. \tag{7.3}$$

The trucks must also match the capacity demand of the mining locations. For each location i, we are only interested in the routes, j, that connect to the location. We denote the set of routes that connect to location i by $J(i)$. Then we have:

$$\sum_{t,m;\, j\in J(i)} P_{t,k,m}\mathbf{f}_{t,j,k,m} \geq D_{i,k} \quad \forall\ k \in \{1, \ldots, K\}, i \in I. \tag{7.4}$$

The capacity constraints must be satisfied with the set of compatible trucks and loaders for each location. That is, from the chosen fleet of trucks we must consider whether the set of trucks that are *compatible* with the loaders are capable of fulfilling the capacity constraints.

Theorem 7.2 *Suppose we model the equipment selection problem as a mixed-integer linear program with a minimising cost objective function. Then production feasibility is not guaranteed with constraints* (7.2) *and* (7.4) *alone.*

Proof Consider one loading location. Let there be exactly two types of loaders operating at this location, λ_1 and λ_2. From constraint (7.2), we have:

$$P_{\lambda_1} + P_{\lambda_2} \geq D.$$

That is, the productivity of the loader fleets of type λ_1 and λ_2 meet the productivity requirements at the location. Suppose we have two types of trucks servicing the location, τ_1 and τ_2. From constraint (7.4) we have:

$$P_{\tau_1} + P_{\tau_2} \geq D.$$

That is, the productivity of the truck fleets of type τ_1 and τ_2 meet the productivity requirements of the location.

Next, suppose that the compatibility sets of loader types with truck types are different for each type of loader. Specifically, loader λ_1 is only compatible with truck τ_1, and loader λ_2 is only compatible with truck τ_2.

Let $P_{\tau_2} = D$ and $P_{\lambda_1} = P_{\lambda_2} = \frac{D}{2}$. Then, since it is a minimisation problem, the constraints (7.2) and (7.4) are met minimally and the actual productivity capability of the trucks and loaders (when working together) is $\frac{D}{2}$ and the productivity requirements of the mine are not met. ■

Therefore, we must ensure that the weak productivity constraint is satisfied for every possible subset of truck and loader fleets. To capture this in a constraint set, we first recall that the set of loader types is denoted by L. Next, we define the set $T(L')$ to be the set of truck types that are compatible with the subset of loader types $L' \subset L$.

Theorem 7.3 *The compatibility of the selected fleets is ensured, in combination with constraint* (7.2)*, by the following superset constraint set:*

$$\sum_{t \in T(L'),m} P_{t,k,m} \mathbf{f}_{t,j,k,m} \geq \sum_{l \in L',m} P_{l,k,m} \mathbf{f}_{l,i,k,m} \qquad \forall\ L' \subset L, k, i. \tag{7.5}$$

Proof Suppose it is not sufficient. Then there exists some combination of equipment such that the compatibility prevents satisfaction of productivity requirements. Let this set of equipment be represented by L^1 and $T(L^1)$. From constraint (7.5) we know that:

$$\sum_{t \in T(L^1),m} P_{t,k,m} \mathbf{f}_{t,j,k,m} \geq \sum_{l \in L^1,m} P_{l,k,m} \mathbf{f}_{l,i,k,m}.$$

That is, the productivity of the set of trucks at least matches the productivity of its compatible loader set. This truck set cannot be the only equipment, otherwise

$$\sum_{t,m} P_{t,k,m} \mathbf{f}_{t,j,k,m} \geq D_{t,k}$$

and we are done. Therefore the selected truck set includes the subset $T(L^1)$ and some other subset $T(L^2)$.

That is, we have

$$\sum_{l \in L^1,m} P_{l,k,m} \mathbf{f}_{l,i,k,m} + \sum_{l \in L^2,m} P_{l,k,m} \mathbf{f}_{l,i,k,m} \geq D_{k,i},$$

for all i, k. That is, we have

$$\begin{aligned} \sum_{\substack{t \in T(L^1),\\ m}} P_{t,k,m} \mathbf{f}_{t,j,k,m} + \sum_{\substack{t \in T(L^2),\\ m}} P_{t,k,m} \mathbf{f}_{t,j,k,m} &\geq \sum_{\substack{l \in L^1,\\ m}} P_{l,k,m} \mathbf{f}_{l,i,k,m} + \sum_{\substack{l \in L^2,\\ m}} P_{l,k,m} \mathbf{f}_{l,i,k,m} \\ &\geq D, \end{aligned}$$

and the requirements are met. ■

Since L' comes from the power set of L, the compatibility constraint set will generate $K|I|(2^{|L|} - 1)$ constraints (where K is the total number of periods and $|I|$ is the total number of locations). As a power set constraint, it should be implemented using a separation algorithm. However, for a given case study the number of loaders possible in the final solution will generally be much lower than the complete set. In this case we can limit the generation of constraints to a maximum of α loader types. This will produce $K|I|\sum_{a=1}^{\alpha}(\frac{|L|!}{a!(|L|-a)!})$ constraints. To further reduce the overall number of these constraints in the solver, we use a separation algorithm. With this branch-and-cut method, we begin with no compatibility constraints in the model. We iteratively solve the model and check for feasibility—any violated constraints are then added into the model before it is resolved.

We link the equipment tracking variables, $\mathbf{x}_{t,k,m}$, to the allocation variables $\mathbf{f}_{t,j,k,m}$ by placing an upper bound on the allocation in the following *coupling constraints* (with a corresponding constraint for loaders):

$$\mathbf{x}_{t,k,m} \geq \sum_{j} \mathbf{f}_{t,j,k,m} \qquad \forall\ t \in T, k \in \{1, \ldots, K\}, m \in \{1, \ldots, M\}. \tag{7.6}$$

We ensure that, in each period, we can only own non-new equipment if we owned it in the previous period, as captured in the following *precedence constraints* (with a corresponding constraint for loaders):

$$\mathbf{x}_{t,k,m} = x_{t,k-1,m-1} - \mathbf{s}_{t,k,m} \qquad \forall t \in T, k \in \{2, \ldots, K\}, m \in \{2, \ldots, M\}, \tag{7.7}$$

$$x_{t,k-1,m-1} \geq \mathbf{s}_{t,k,m} \qquad \forall t \in T, k \in \{2, \ldots, K\}, m \in \{2, \ldots, M\}. \tag{7.8}$$

In this model we consider pre-existing equipment. We only need to consider pre-existing trucks and loaders which are drawn from the subset $\mathbf{P} \subset T \cup L$. Recall that $b(1)$ is the starting age of the pre-existing truck type t. Then, if $\bar{x}_{t,1,b(1)}$ is the number of pre-existing truck of type t, with age $b(1)$, we have (with a corresponding constraint for loaders):

$$x_{t,1,b(1)} + s_{t,1,b(1)} = \bar{x}_{t,1,b(1)} \qquad \forall\ t \in \mathbf{P}. \tag{7.9}$$

7.2.5 Complete Model

$$\begin{aligned} \min \quad & \sum_{t,k} F_t D_k^1 \mathbf{x}_{t,k,1} + \sum_{l,k} F_t D_k^1 \mathbf{y}_{l,k,1} + \sum_{t,k,m,h} B_{t,k,m}^h D_k^1 V_{t,k,b(k)+h-1} \mathbf{f}_{t,j,k,m} \\ & + \sum_{l,k,m,h} B_{t,k,m}^h D_k^1 V_{l,k,b(k)+h-1} \mathbf{f}_{l,i,k,m} - \sum_{t,k,m} F_t D_{k,m}^2 \mathbf{s}_{t,k,m} - \sum_{l,k,m} F_l D_{k,m}^2 \mathbf{s}_{l,k,m} \end{aligned} \tag{7.10}$$

$$
\begin{aligned}
\text{s.t.} \quad & \sum_{l,m} P_{l,k,m}\mathbf{f}_{l,i,k,m} \geq D_{i,k} && \forall\, i, k, && (7.11)\\
& \sum_{t,m;\, j\in J(i)} P_{t,k,m}\mathbf{f}_{t,j,k,m} \geq D_{i,k} && \forall\, i, k, && (7.12)\\
& \sum_{t\in T(L'),m} P_{t,k,m}\mathbf{f}_{t,j,k,m} \geq \sum_{l\in L',m} P_{l,k,m}\mathbf{f}_{l,i,k,m} && \forall\; L' \subset L, i, k, && (7.13)\\
& \mathbf{x}_{t,k,m} \geq \sum_{j} \mathbf{f}_{t,j,k,m} && \forall\, t, k, m, && (7.14)\\
& \mathbf{y}_{l,k,m} \geq \sum_{i} \mathbf{f}_{l,i,k,m} && \forall\, l, k, m, && (7.15)\\
& \mathbf{x}_{t,k,m} = x_{t,k-1,m-1} - \mathbf{s}_{t,k,m} && \forall\, t, k > 1, m > 1, && (7.16)\\
& x_{t,k-1,m-1} \geq \mathbf{s}_{t,k,m} && \forall\, t, k > 1, m > 1, && (7.17)\\
& \mathbf{y}_{l,k,m} = y_{t,k-1,m-1} - \mathbf{s}_{l,k,m} && \forall\, l, k > 1, m > 1, && (7.18)\\
& y_{t,k-1,m-1} \geq \mathbf{s}_{l,k,m} && \forall\, l, k > 1, m > 1, && (7.19)\\
& x_{t,1,b(1)} + s_{t,1,b(1)} = \bar{x}_{t,1,b(1)} && \forall\, t \in \mathbf{P}, && (7.20)\\
& \mathbf{x}_{t,k,m}, \mathbf{y}_{l,k,m}, \mathbf{s}_{t,k,m}, \mathbf{s}_{l,k,m} \in \mathbb{Z}^+, \\
& \mathbf{f}_{t,j,k,m}, \mathbf{f}_{l,i,k,m} \in \mathscr{R}^+.
\end{aligned}
$$

7.3 Computational Study

We consider two case studies as presented in Chap. 5. The first is for a planned mine with no comparative solution. The second is for a mine with pre-existing equipment and a comparative solution. These case studies illustrate that the difficulty in solving the model is not wholly influenced by the size of K or the number of routes in the mine plan. In both case studies, we deal with mine progression over time by allocating a cycle time of zero to those routes that are inaccessible in a given time period.

7.3.1 *Few Locations Case Study Results*

We implemented this case study on a Pentium 4 PC with 3.0 GHz and 2.5GB of RAM. The model was implemented in C++ using *Ilog Concert Technology v 2.5* objects and *Ilog Cplex v 11.0* libraries to solve the problem with default settings. The constraints (7.13) were implemented as lazy constraints—i.e., we separated the constraints from the model and only added those which were violated by feasible MIP solutions. We implemented this problem with 14502 variables (3672 integers) and 6858 constraints. The separation algorithm added a further 2380 compatibility constraints. We ran the 9-period problem for 34 h before the computer memory was exhausted and achieved

Table 7.1 The results summary for the first case study solutions with varying periods

Periods	Variables	Constraints	Time (seconds)	Quality	Solution
7	9100	4242 + 2044	5331	Optimal	1.88599×10^7
8	11648	5473 + 1484	12049	Optimal	1.97785×10^7
9	14502	6858 + 2380	19477	3%	2.05244×10^7

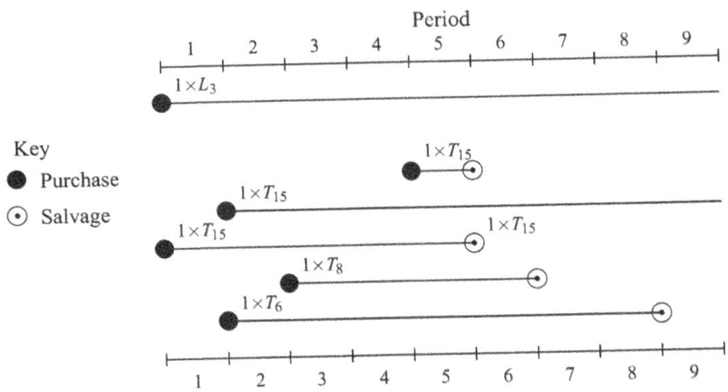

Fig. 7.6 The first case study 9-period purchase and salvage policy with depreciation 50%

a solution within 3% of optimality. The computation times for single solves of the problem with 7, 8 and 9 periods is presented in Table 7.1. The solution obtained is given in Fig. 7.6. In this figure, a black circle depicts a purchase, while an empty circle depicts a salvage. For this case study, the optimal solution was to purchase one loader of type three, and operate these for the entire nine periods. Only five trucks were purchased for the entire 9 periods. The purchasing pattern is indicative of the dramatic increase in production requirements in the first few periods—in the beginning, one truck is sufficient. In reality, this would not be a suitable solution, because if that truck is in maintenance, the mine no longer produces. Also, it means that the loader is not utilised while the truck is delivering its load. In the subsequent two periods, three further trucks are purchased. By period four, there are three types of trucks working in the fleet. In period five, the productivity of the existing trucks has fallen sufficiently so that a new purchase is worthwhile. The fifth truck is then salvaged only 1 year later. While is it not realistic that the truck is sold after only one year, it is conceivable that the truck is relocated to another mine in need of a relatively new truck. The remaining trucks are gradually salvaged as the needs of the mine decrease, leaving only one truck until the end of the final period.

The truck allocation solution is presented in Table 7.2. It is difficult to identify the slack in the allocation of equipment because the model did not motivate a minimum utilisation value—i.e., it does not cost more (in terms of the objective function value) to allocate trucks to locations for more time than necessary. Some routes receive every

Table 7.2 The 9-period truck allocation policy for the *few locations* case study, 50% depreciation

Route	Period								
	1	2	3	4	5	6	7	8	9
(a)	0.01 $T_{15}(1)$	0.14 $T_{15}(2)$	1.00 $T_6(2)$	0.44 $T_{15}(4)$	0.51 $T_8(3)$	0.51 $T_{15}(5)$	0.62 $T_{15}(6)$	0.89 $T_{15}(8)$	0.11 $T_{15}(8)$ 1.00 $T_{15}(9)$
(b)	0.99 $T_{15}(1)$	0.97 $T_6(1)$	0.58 $T_8(1)$	0.14 $T_{15}(3)$ 0.56 $T_{15}(4)$	0.88 $T_6(4)$	1.00 $T_8(4)$ 0.26 $T_{15}(5)$	1.00 $T_6(6)$ 0.74 $T_{15}(7)$	0.07 $T_6(7)$ 1.00 $T_{15}(7)$ 0.11 $T_{15}(8)$	0.74 $T_{15}(8)$
(c)			0.23 $T_{15}(2)$	0.24 $T_{15}(3)$	0.07 $T_8(3)$	0.23 $T_{15}(5)$ 0.21 $T_{15}(6)$	0.38 $T_{15}(6)$	0.19 $T_6(7)$	0.10 $T_{15}(8)$
(d)		0.03 $T_6(1)$ 1.00 $T_{15}(1)$ 0.86 $T_{15}(2)$	0.42 $T_8(1)$ 0.77 $T_{15}(2)$ 1.00 $T_{15}(3)$	0.62 $T_{15}(3)$ 1.00 $T_8(2)$ 0.44 $T_9(3)$	0.12 $T_6(4)$ 0.41 $T_8(3)$ 1.00 $T_{15}(1)$ 1.00 $T_{15}(4)$ 1.00 $T_{15}(5)$	1.00 $T_6(5)$ 0.79 $T_{15}(6)$	0.26 $T_{15}(7)$	0.73 $T_6(7)$	0.06 $T_{15}(8)$

Key $x\,T_t(m)$ indicates that x trucks of type t and age m operate on the route in the given period.

Table 7.3 The 9-period loader allocation policy for the first case study, 50% depreciation

Route	Period								
	1	2	3	4	5	6	7	8	9
(a)	0.01 $L_3(1)$	0.03 $L_3(2)$	0.07 $L_3(3)$	0.06 $L_3(4)$	0.09 $L_3(5)$	0.07 $L_3(6)$	0.08 $L_3(7)$	0.09 $L_3(8)$	0.10 $L_3(9)$
(b)	0.09 $L_3(1)$	0.36 $L_3(2)$	0.30 $L_3(3)$	0.24 $L_3(4)$	0.23 $L_3(5)$	0.33 $L_3(6)$	0.19 $L_3(7)$	0.20 $L_3(8)$	0.08 $L_3(9)$
(c)			0.05 $L_3(3)$	0.05 $L_3(4)$	0.01 $L_3(5)$	0.04 $L_3(6)$	0.05 $L_3(7)$	0.02 $L_3(8)$	0.01 $L_3(9)$
(d)		0.38 $L_3(2)$	0.39 $L_3(3)$	0.44 $L_3(4)$	0.47 $L_3(5)$	0.17 $L_3(6)$	0.02 $L_3(7)$	0.01 $L_3(8)$	0.01 $L_3(9)$

Key $x\,L_l(m)$ indicates that x loaders of type l and age m operate on the route in the given period.

Table 7.4 The results summary for the second case study with depreciation 50% and 13 periods

Periods	Variables	Constraints	Time (seconds)	Quality (%)	Objective function value($)
10	39855	9814 + 3319	5643	3	1.26292×10^8
11	47166	11599 + 3043	3979	3	1.31263×10^8
12	55032	13521 + 4043	17656	3	1.37168×10^8
13	63433	15571 + 3795	26662	3	1.37249×10^8

type of truck in the fleet. This could have an impact on cycle time, as some trucks will be faster than others or will be served faster at the loader. However, this level of detail is not captured in our model. The loader allocation solution splits one loader across two mining locations [Table 7.3]. In some mining scenarios where loader movement across the mine is prevented, this solution would be unrealistic. It is likely that one loader at each location is preferred. In this case, it is easy to add constraints that reflect node-disjointed flow for the loaders—for example, this can be achieved by forcing $\mathbf{f}_{l,i,k,m} \in \mathbb{Z}$.

This case study seems small with respect to the number of periods and the number of routes and locations. However, the symmetry in this problem—arising because of the number of identical equipment which can be allocated to the same decisions—makes it difficult to solve with default settings. We will defer comparisons with the second case study to the Discussion Sect. 7.3.3.

7.3.2 *Many Locations Case Study Results*

Our industry partner provided a second case study from an ongoing mining operation with pre-existing equipment and a more complex route structure. This case study considers a surface mine operating under a truck-loader hauling system, and mines ore and waste in an open pit.

We implemented this case study on a Pentium 4 PC with 3.0 GHz and 2.5GB of RAM. The model was implemented in C++ using *Ilog Concert Technology v 2.5* objects and *Ilog Cplex v 11.0* libraries to solve the problem. The mixed-integer program contained 63433 variables (5304 integer) and 19366 constraints; we set up the 13-period problem with just 15571 constraints before the compatibility constraints were taken into account.

Table 7.4 shows the number of constraints that were added by the separation algorithm overall for the full problem and versions of the problem with a reduced schedule length (namely, 10, 11 and 12 years).

After 7.5 h of algorithm run-time, we obtained a solution within 3% of the optimal solution for the 50% depreciation case study. When the algorithm was permitted to run for a longer period, the computer memory was exhausted. However, this is a satisfactory optimality gap for this application, as we will illustrate later with a retrospective solution comparison.

The purchase and salvage policy for this multi-location mine is complicated by the capacity requirements, which contain several significant changes from period to period. This leads to short-term ownership of some trucks. For example a type-8 truck was purchased in period 8 and salvaged at the start of period 9. The complete purchase and salvage policy is sketched in Fig. 7.7, where pre-existing equipment is indicated by the P index.

The allocation policy is shown in Tables 7.5, 7.6, 7.7 and 7.8. In the allocation policy for this case study, we represent the age of the equipment in parentheses as an equipment tracking tool. Since the age of the equipment is a factor in the cost of operating the equipment, it is important to allocate the correct age equipment as dictated by the policy.

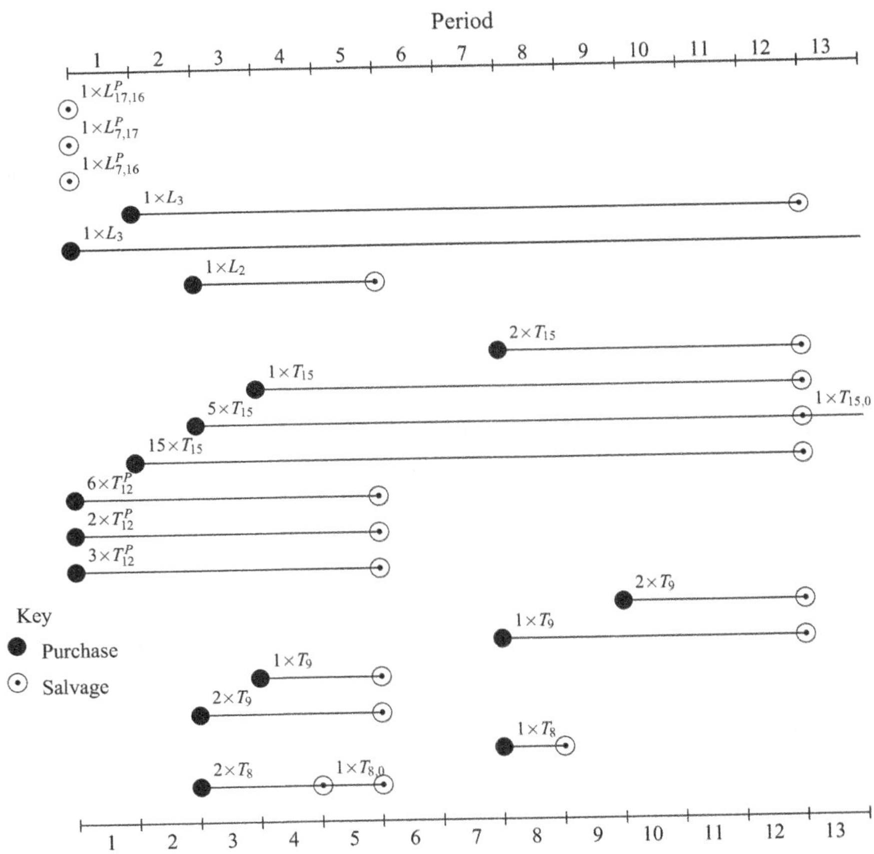

Fig. 7.7 The 3% optimal 13-period solution for case study two with the depreciation parameter set to 50%

For the purposes of mine management, it would be a simple task to create a spreadsheet that can reflect the flexibilities in the policy and allow dynamic changes in the policy without affecting the objective function value.

7.3.3 Discussion

In the final solutions for the case studies, the model has selected 3 and 4 truck types respectively. However, it is considered unusual in industry to have more than three types of trucks, and generally this number would only arise due to pre-existing equipment. This suggests that the models are not reflecting the true penalties associated with the fixed costs of owning different types of equipment, or conversely, that the reasoning behind homogeneous or small mixes of fleet needs further justification. Also, operating different truck types on the same route may influence the accuracy of the cycle times due to bunching of equipment. This issue is usually addressed during dispatching of equipment, but ideally should be considered during equipment selection—i.e., to account for the interactive effect of the selected equipment. It is not obvious how to incorporate this into the current deterministic mixed-integer program, but it makes an interesting question for future research.

As some of the data for the case studies was limited or unknown (such as the requirement for stockpiles to have their own loaders, or when locations are mined simultaneously within a period) the solutions generated by our model sometimes requires that one loader move from one location to another. This may not be realistic and in this case can be amended by enforcing integrality constraints on the loader decision variables.

The first case study had many less routes than the first, which made the overall size of the problem significantly smaller. However this problem was still difficult to solve, demonstrating the difficulty in differentiating between similar pieces of equipment over long-term schedules. The problem exhibits a lot of symmetry, which, if addressed, would lead to faster solving times.

One form of validation is to compare our solutions with the actual solution implemented in the mine. We are fortunate to have obtained this information for the second of the two case studies. We do not know the complete process behind the derivation of the industry solutions we present here. However, from discussions with our partner combined with our knowledge of the literature, we will describe our understanding of the process.

For the second case study, three equipment selection alternative solutions were given to us. The first and cheapest solution was created on an in-house equipment selection spreadsheet. This spreadsheet was not designed to consider mixed fleets—i.e. all trucks selected must be compatible with all loaders. Furthermore, the truck fleet must be homogeneous. The spreadsheet functioned by first selecting the loaders by minimising the cost of operation such that the capacity constraints were met. Then a *match factor equation* (such as that discussed in [3]) was used to determine the best fleet size for trucks, whose type was pre-determined by the pre-existing truck fleet.

Table 7.5 The truck allocation policy for case study two with 13 periods and 50% depreciation (first 7 periods)

Routes	Period 1	2	3	4	5	6	7
(a)	1.81 $T_{12}(8)$ 6.00 $T_{12}(12)$						
(b)	0.35 $T_{12}(8)$						
(c)	0.17 $T_{12}(8)$	9.42 $T_{15}(1)$	6.00 $T_{12}(14)$ 0.97 $T_{15}(1)$				
(d)	2.00 $T_{12}(9)$	5.90 $T_{12}(13)$	0.68 $T_{12}(10)$ 2.00 $T_{12}(11)$				
(e)	0.05 $T_{12}(8)$	0.18 $T_{12}(9)$ 0.09 $T_{12}(13)$	3.70 $T_{15}(1)$	2.65 $T_9(2)$ 1.50 $T_8(2)$ 6.73 $T_{15}(3)$	0.42 $T_{12}(13)$ 6.00 $T_{12}(16)$		
(f)	0.61 $T_{12}(8)$	2.32 $T_{12}(9)$ 2.00 $T_{12}(10)$ 4.87 $T_{15}(1)$	2.00 $T_9(1)$ 2.32 $T_{12}(10)$ 0.32 $T_{15}(1)$ 15.0 $T_{15}(2)$ 0.99 $T_8(1)$	1.00 $T_9(1)$ 2.00 $T_9(2)$ 3.00 $T_{12}(11)$ 1.97 $T_{12}(12)$ 6.00 $T_{12}(15)$	3.00 $T_{12}(12)$ 1.58 $T_{12}(13)$ 1.00 $T_{15}(2)$ 5.00 $T_{15}(3)$ 0.67 $T_{15}(4)$		
(g)		0.31 $T_{15}(1)$					
(h)		0.40 $T_{15}(1)$	1.00 $T_8(1)$				
(i)		0.50 $T_{12}(9)$					
(j)				0.50 $T_8(2)$ 0.03 $T_{12}(12)$	2.67 $T_{15}(4)$	12.84 $T_{15}(5)$	13.115 $T_{15}(6)$
(k)				1.00 $T_{15}(1)$ 5.00 $T_{15}(2)$ 8.27 $T_{15}(3)$	11.7 $T_{15}(4)$	1.00 $T_{15}(3)$ 4.26 $T_{15}(4)$ 2.16 $T_{15}(5)$	1.00 $T_{15}(4)$ 5.00 $T_{15}(5)$ 1.59 $T_{15}(6)$
(l)						0.74 $T_{15}(4)$	
(m)							0.30 $T_{15}(6)$

Key $x\,T_t(m)$ indicates that x trucks of type t and age m operate on the route in the given period.

Table 7.6 The truck allocation policy for the second case study solution with 13-periods and 50% depreciation (last 6 periods)

Routes	Period					
	8	9	10	11	12	13
(a)						
(b)						
(c)						
(d)						
(e)		1.00 $T_9(2)$ 15.00 $T_{15}(8)$	1.00 $T_{15}(7)$ 5.00 $T_{15}(8)$ 6.85 $T_{15}(9)$	2.00 $T_9(2)$ 1.00 $T_9(4)$ 2.00 $T_{15}(4)$ 1.00 $T_{15}(8)$ 5.00 $T_{15}(9)$ 0.83 $T_{15}(10)$	2.00 $T_9(3)$ 1.00 $T_9(5)$ 0.72 $T_{15}(5)$ 1.00 $T_{15}(9)$	0.01 $T_{15}(11)$
(f)		2.00 $T_{15}(2)$ 1.00 $T_{15}(6)$ 5.00 $T_{15}(7)$	2.00 $T_9(1)$ 1.00 $T_9(3)$ 2.00 $T_{15}(3)$ 8.15 $T_{15}(9)$	14.2 $T_{15}(10)$	6.16 $T_{15}(11)$	0.99 $T_{15}(11)$
(g)						
(h)						
(i)						
(j)	1.00 $T_8(1)$ 1.00 $T_9(1)$ 10.6 $T_{15}(7)$				1.28 $T_{15}(5)$ 6.43 $T_{15}(11)$	
(k)	2.00 $T_{15}(1)$ 1.00 $T_{15}(5)$ 5.00 $T_{15}(6)$ 4.37 $T_{15}(7)$				5.00 $T_{15}(10)$ 2.41 $T_{15}(11)$	
(l)						
(m)						

Key $xT_t(m)$ indicates that x trucks of type t and age m operate on the route in the given period.

This solution kept one $34T$ loader but salvaged the other two pre-existing loaders. Two new $40T$ loaders were purchased to meet the capacity constraints. The existing truck fleet was kept, and further trucks (of the same type) were added in period one (five trucks), period two (two trucks) and period five (one truck). This selection policy cost $\$1.51483 \times 10^8$ (using our objective function) for the full 13 period schedule. In comparison, our policy yielded a saving of $\$14,234,000$—a saving of 9.4%.

The second solution provided by the industry partner contained the same truck purchase and salvage policy. However, an equipment selection manager believed that all loaders should be salvaged, and that a new $42T$ and two $57T$ loaders should be purchased instead. In this solution, the choice of loaders is restricted in that they must be compatible with the existing truck fleet. This policy yielded a cost of $\$1.55241 \times 10^8$. Comparatively, our solution saved $\$17,992,000$ or 11.6%.

Table 7.7 The loader allocation policy for the second case study with 13 periods and 50% depreciation (first 7 periods)

Locations	Period						
	1	2	3	4	5	6	7
M1	0.13 $L_3(1)$						
M2	0.04 $L_3(1)$	0.70 $L_3(1)$	0.43 $L_3(3)$				
M3	0.12 $L_3(1)$	0.21 $L_3(1)$ 0.76 $L_3(2)$	0.51 $L_2(1)$ 1.00 $L_3(2)$ 0.56 $L_3(3)$	1.00 $L_2(2)$ 0.93 $L_3(4)$	0.33 $L_2(3)$ 1.00 $L_3(4)$		
M4		0.26 $L_3(1)$					
S1		0.01 $L_3(1)$	0.05 $L_2(1)$				
S2		0.05 $L_3(1)$					
S3				0.73 $L_3(3)$	0.67 $L_2(3)$ 0.68 $L_3(5)$	0.93 $L_3(5)$ 0.13 $L_3(6)$	0.89 $L_3(6)$
S4							0.03 $L_3(6)$

Key $x L_l(m)$ indicates that x loaders of type l and age m operate on the route in the given period.

Table 7.8 The loader allocation policy for the second case study with 13 periods and 50% depreciation (last 6 periods)

Locations	Period					
	8	9	10	11	12	13
M1						
M2						
M3		0.34 $L_3(8)$ 0.97 $L_3(9)$	0.38 $L_3(9)$ 1.00 $L_3(10)$	0.26 $L_3(10)$ 1.00 $L_3(11)$	0.47 $L_3(12)$	0.01 $L_3(13)$
M4						
S1						
S2						
S3	0.13 $L_3(7)$ 1.00 $L_3(8)$				0.07 $L_3(11)$ 0.53 $L_3(12)$	
S4						

Key $x L_l(m)$ indicates that x loaders of type l and age m operate on the route in the given period.

Finally, the third solution purchased a loader fleet that maintained full homogeneity—three 57T loaders. The truck purchase and salvage policy was the same as above. This was the actual solution adopted for the mine, and cost \$1.66550 $\times$ 10^8. Comparatively our solution saved \$29, 301, 000 or 17.6%.

It is interesting to observe that the industry partner selected the most expensive of their three solutions. This indicates that they attribute a value to having homogeneous fleets. Since we did not account for the cost of ancillary equipment, on-costs for spares and the training of artisans for maintenance and equipment use, it is difficult to make a true comparison of these solutions. However, it is clear that there is an advantage to using an integer programming approach if all the relevant data is available.

7.4 Conclusion

In this chapter, we have captured the truck and loader equipment selection problem for surface mines in a large-scale mixed-integer program, developing preprocessing and adopting a separation algorithm to improve the tractability. We used a continuous allocation variable that suggested the optimal portion of the fleet that should work at each location. These variables created a flexible allocation policy alongside the purchase and salvage policy generated by the model. This is a useful tool for mining engineers, who may take the allocation policy as further evidence that the selected fleet would be able to perform the required tasks under uncertainty, or simply use the policy as a guide to manage the fleet.

In our formulation, we paid particular attention to reducing the discretisation error in the objective function. We provided a formula that accurately accounts for cost shifts that occur when equipment moves from one age bracket to the next within a period. We solved two case studies over the entire mine plan, demonstrating the model is computationally effective for real world problems.

One important assumption is that the mine plan is deterministic. However, the mine plan is dependent on future demand. Ideally, the equipment selection model would be robust to uncertainty in the market and other stochastic elements, such as machine breakdown, maintenance and truck cycle times. Modelled as a stochastic mixed-integer program, this would decrease the tractability of the model as it stands. However, capturing the uncertainty, stochasticity and interactive effects of truck types are important starting points for future research on this topic.

Notation Index

A summary of the notation used in this chapter is as follows.

- B_0 the size of an age bracket.
- H_k the size of a period.
- I the set of all mining locations.
- J the set of all dumpsites.
- T the set of all truck types.
- L the set of all loader types.
- K the total number of periods.
- t the truck type index, $t \in T$.
- l the loader type index, $l \in L$.
- i the mining location index.
- j the route index.
- m the equipment age bracket index.
- k the period index.
- $\mathbf{x}_{t,k,m}$ number of trucks of type t owned in period k which are in age bracket m (integer variable).
- $\mathbf{y}_{l,k,m}$ number of loaders of type l owned in period k which are in age bracket m (integer variable).

$\mathbf{s}_{t,k,m}$ number of trucks of type t salvaged in period k which are in age bracket m (integer variable).
$\mathbf{s}_{l,k,m}$ number of loaders of type l salvaged in period k which are in age bracket m (integer variable).
$\mathbf{f}_{t,j,k,m}$ portion of trucks of type t, in age bracket m, that are allocated to route j in period k, where $\mathbf{f}_{t,j,k,m} \in [0, \mathbf{x}_{t,k,m}]$ (continuous variable).
$\mathbf{f}_{l,i,k,m}$ portion of loaders of type l, in age bracket m, that are allocated to location i in period k, where $\mathbf{f}_{l,i,k,m} \in [0, \mathbf{y}_{l,k,m}]$ (continuous variable).
$M_t^{\max}$ the maximum age bracket of truck type t.
$M_k(t)$ the maximum age bracket of truck type t in time period k.
$A_{t,m}$ the availability of truck type t in age bracket m.
C_t the capacity of truck t.
$\tau_{t,j}$ the cycle time of truck t on route j.
F_t the fixed cost of purchasing truck type t.
I a fixed interest rate used for net present value.
$\beta(k')$ the base of the equipment ageing recursion.
$b(k)$ the age bracket in which equipment lies at the beginning of a period k.
h an indicator selecting the case for age bracket.
$V_{t,k,b(k)+h-1}$ the variable cost of truck type t in period k while in age bracket $b(k) + h - 1$.
J the depreciation rate.
$D_{i,k}$ the demand of location i in period k.
$P_{l,k,m}$ the maximum possible productivity of loader l, aged m in period k.
$\mathbf{P}$ the set of pre-existing equipment.

Acknowledgements Components of this chapter were presented in [1, 2].

References

1. C. Burt, An optimisation approach to materials handling in surface mines. Ph.D. thesis, Mathematics and Statistics, Curtin University of Technology, Bentley, Perth, Australia, 2008. http://espace.library.curtin.edu.au:80/R?func=dbin-jump-full&local_base=gen01-era02&object_id=165737
2. C.N. Burt, L. Caccetta, L. Fouché, P. Welgama, An MILP approach to multi-location, multi-period equipment selection for surface mining with case studies. J. Ind. Manag. Optim. **12**(2), 403–430 (2016)
3. C.N. Burt, L. Caccetta, Match factor for heterogeneous truck and loader fleets. Int. J. Min. Reclam. Environ. **21**, 262–270 (2007)
4. E. Topal, S. Ramazan, A new MIP model for mine equipment scheduling by minimizing maintenance cost. Eur. J. Oper. Res. **207**(2), 1065–1071 (2010)

Chapter 8
Utilisation-Based Equipment Selection

Christina N. Burt, Louis Caccetta and Yao-ban Chan

8.1 Introduction

In a surface mining operation, the operating cost of equipment is often represented by a *cost per hour*. The cost of running equipment may change with the age of the equipment, usually reflecting maintenance expenses (Fig. 8.1). When accounting for cost in an equipment selection model, we are thus interested in determining the **utilisation** and cumulative utilisation (or the equipment age) so that we may best account for the cost of the equipment.

The models presented in Chaps. 6 and 7 have adhered to the assumption of *full period utilisation* (Sect. 6.2.1). By this assumption, if equipment is owned in a particular period then we assume that it was utilised to the fullest possible extent in this period. Clearly it is not ideal to be charging full utilisation if the equipment is not operating to capacity—we will favour the selection of equipment that is slightly cheaper to run for the entire period but not necessarily cheaper to run if charged by utilisation. In addition, in previous chapters this assumption may have been responsible for the selection of multiple truck types in the optimal solutions. This may have

---

The original version of this chapter was revised: Originator type has been changed and the chapter author names have been included. The erratum to this chapter is available at https://doi.org/10.1007/978-3-319-76255-5_11

C. N. Burt (✉)
Department of Mathematics and Statistics, The University of Melbourne, Parkville, VIC, Australia
e-mail: c.burt@ms.unimelb.edu.au

L. Caccetta
Department of Mathematics and Statistics, Curtin University, Bentley, WA, Australia
e-mail: L.Caccetta@curtin.edu.au

Y.-b. Chan
The University of Melbourne, Melbourne, VIC, Australia
e-mail: Leon.Fouche@riotinto.com

C. N. Burt and L. Caccetta (eds.), *Equipment Selection for Mining: With Case Studies*, Studies in Systems, Decision and Control 150,
https://doi.org/10.1007/978-3-319-76255-5_8

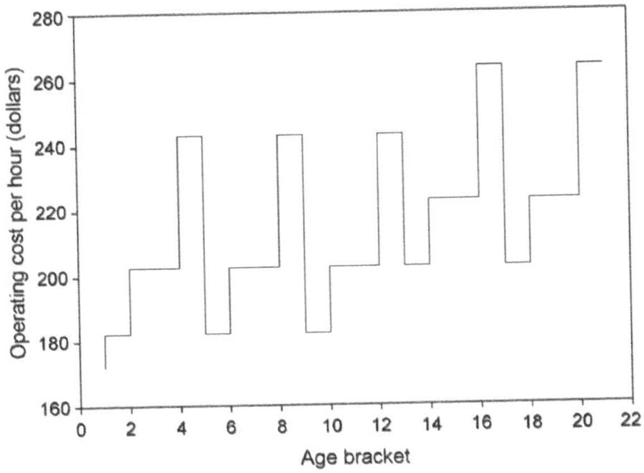

Fig. 8.1 The variance of equipment operating cost against cost bracket. The rise in operating cost can reflect the increased maintenance expense or the time since the last overhaul. Large drops in the operating expense occur when an overhaul has taken place

been caused by the discrepancy between the fleet productivity levels and the required productivity levels: the optimiser would rather select a small cheap truck to fill a gap than pay for a full period of a more expensive but larger truck.

In this chapter, we discard the assumption of full period utilisation and keep track of how much each equipment is actually utilised. This allows us to bring a utilisation variable into the objective function, thus preventing overstatement of operating cost. By doing this we can:

- estimate salvage value for the current period;
- track how close the equipment is to its enforced retirement age; and
- maintain the equipment according to its maintenance schedule.

However, the mine requirements may change from period to period, so the utilisation of any piece of equipment will change too. This means that the cumulative utilisation of a piece of equipment may be a non-uniform piecewise linear function. Thus our challenge is:

To account for utilisation when it is a non-uniform piecewise linear function.

As equipment is discrete, it is appropriate to use integer variables to keep track of the purchase and salvage of trucks and loaders. Ideally, we would therefore like to express all our variables as integers to take advantage of algorithms available to solve a pure integer program. In this vein, we could represent the utilisation of equipment by an integer variable containing the number of hours that equipment has worked. However, it is more natural to formulate utilisation as the proportion of total available time spent working, which would require a continuous variable.

We know from previous chapters that the production constraints are naturally linear and that the cost objective function can be linearised through careful definition of the variable indexes. Therefore, the equipment selection problem with utilisation cost objective can be best expressed as a mixed-integer linear programming model. To help us deal with the complexities of this model, we only consider a single-location mine (Fig. 8.2). We represent a single-location mine by a loading location connected to a dump-site by a single truck route.

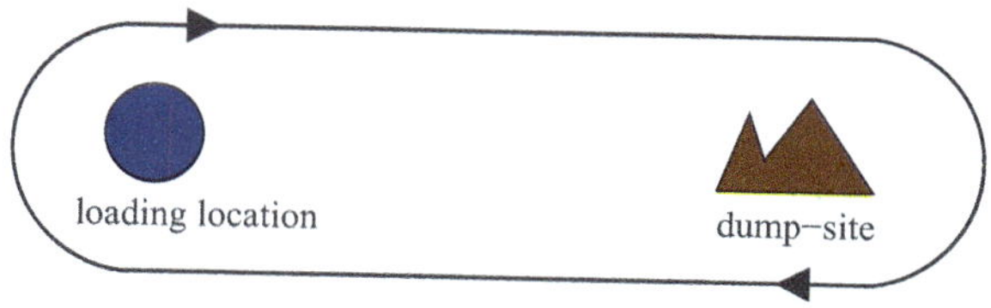

Fig. 8.2 A single-location mine model

In this chapter, we:

- Consider the inclusion of pre-existing equipment and possibly heterogeneous fleets. If equipment selection is performed part-way through the mining schedule, then a pre-existing fleet must be considered and we must allow the tool to select different types of equipment if the current equipment is obsolete. Neither pre-existing equipment nor heterogeneous fleets have been previously considered in a surface mining equipment selection optimisation model.
- Consider a multiple-period mining schedule. The production requirements of the mine and the truck cycle time can both change significantly over time, and we wish to optimise the selected fleet over all time periods rather than considering each period individually. Although this seems like an obvious consideration for optimisation, multiple-period models are not common in the mining literature.
- Introduce a set of linear integer constraints that account for non-uniform piecewise linear ageing of the equipment.
- Present and solve a new mixed-integer linear program for multi-period, single-location surface mine equipment selection that accounts for equipment utilisation in the cost function. The solution presents a purchase and salvage policy, and also an optimised utilisation policy.

We formalise the model in Sect. 8.2. First, we list the assumptions associated with our solution (Sect. 8.2.1). We then describe the decision variables (Sect. 8.2.2) before deriving the objective function (Sect. 8.2.3) and constraints (Sect. 8.2.4). We provide the complete model (Sect. 8.2.5). To test the model, we consider a surface mining case study in Sect. 8.2.5 that we solve to optimality over four periods. Finally, we discuss opportunities for extending this work in Sect. 8.6.

8.2 The Model

8.2.1 Assumptions

For this formulation, the following assumptions apply:

- Known mine schedule—We presume that an acceptable mine schedule has already been derived, and that the mining method has been selected. We start the equipment selection with a sub-set of trucks and loaders that suit the particular mining scenario.

- Single mining location—For this model, all the loaders and trucks are considered to operate as one fleet. That is, all loaders work in the same location, and all trucks service all the loaders. We consider the mine to have a single mining location, a single dump-site and a single haul route connecting the two.
- Salvage—All equipment is salvageable at some depreciated value of the original capital expense. Any pre-existing equipment may be salvaged at the start of the first time period.
- No auxiliary equipment—Auxiliary equipment, such as wheel loaders and small trucks, are not considered in this model. Although the cost of running auxiliary equipment may differ according to the overall fleet selection, for the purpose of this model this cost is considered trivial. Note that the cost of auxiliary equipment can be built into the operating cost if necessary.
- Known operating hours—The operating hours of the mine are estimated by taking planned downtime, blasting and weather delays into account.
- Single truck cycle time—Since we assume a single mining location, an average truck cycle time is used for all trucks. The cycle time is constant over a period and is known for all periods. It accounts for factors that affect the truck performance, such as rimpull, rolling resistance, haul distance and haul grade.
- Heterogeneous fleets—We allow different types of equipment to work side-by-side.
- Fleet retention—All equipment is retained at the end of the last period. Optionally, we can relax this assumption and allow the model to salvage some or all equipment at the end of the final period.
- Age brackets—We count the age of the equipment in terms of hours utilised. To reduce the number of variables, and to bring the variable structure in line with industry standards, we divide the age into brackets.

8.2.2 *Decision Variables*

In this chapter, we denote the set of all truck types by $\mathbf{X}$ and the set of all loader types by $\mathbf{X}'$. We use i and i' to denote a single truck and loader type, respectively. To simplify the notation, we use e to represent a single type (which can be a truck or loader type), where $e \in \mathbf{X} \bigcup \mathbf{X}'$.

8.2.2.1 Tracking Variables and Fleet Purchase

We define the following binary variables for tracking the equipment:

$\mathbf{x}_{e,j}^{k,l}$: 0-1 selection of one truck or loader of type e with identification number j in time period k and age bracket l.

We use the index j to track individual equipment, which is necessary in order to calculate individual equipment age.

In the first period, we can examine the tracking variable for $k = 0$, $l = 0$ to see if a purchase has been made. However, a different term is required for purchases in time periods $k > 0$. This is derived in Sect. 8.2.3.

We would like to use the index l to denote the age of the equipment, in periods. However, this is impossible, as the equipment age depends on the number of hours the equipment has been utilised, rather than the amount of time since its purchase. One way to do this is by letting l denote utilisation in hours. However, this would lead to a large number of unused variables. Since, in the mining industry, the equipment costings are typically divided into age brackets (such as 5000 h brackets), we can let l denote the utilisation, counted in age brackets. This reduces the number of variables in the model. The utilisation is permitted to change each period, creating a non-uniform piecewise linear ageing function. As we will see, this complicates the variable transition constraints if we wish to retain linearity in them.

Note that we do not count different age brackets in a single time period; only the tracking variable with the earliest age bracket is set to 1 for that period.

8.2.2.2 Fleet Salvage Variables

The identification index j is also necessary for the salvage variables:

$\mathbf{s}_{e,j}^{k,l}$: 0-1 indicator for when the truck or loader with identification number j, of type e, is salvaged in period k while in age bracket l.

Note that the equipment operates for another period before salvage can take place. Consequently, the equipment may move into the next age bracket as the salvage takes place. Alternatively it may be salvaged from the current age bracket. We can cover both cases by saying that salvage occurs in time period k when $\mathbf{x}_{e,j}^{k-1,l} - \mathbf{x}_{e,j}^{k,l} - \mathbf{x}_{e,j}^{k,l+1} = 1$. For pre-existing equipment only, a salvage at the start of the schedule occurs when $\mathbf{x}_{e,j}^{0,l} = 0$.

8.2.2.3 Equipment Utilisation Variables

We define a continuous variable to represent the utilisation of the equipment, in terms of proportion of available time.

$\mathbf{f}_{e,j}^{k,l}$: utilised proportion of total time that the truck or loader identified by j, of type e, in age bracket l will work in time period k.

8.2.3 Objective Function

We wish to minimise the cost of materials handling. The cost comes from two sources: capital expense of the fleet, and operating costs. Offset against this is the revenue gained by salvaging equipment.

First we consider the capital expense, which is a one-off cost at the start of the period. To find out when we have purchased a truck or loader, we look at the tracking variables. For the case $k = 0$ this is straightforward—we simply look at the variable $\mathbf{x}_{e,j}^{0,0}$ (while accounting for any pre-existing equipment that may happen to fall into that age-bracket). However, for $k > 0$ the use of this term may result in overcounting. This is because it is likely for equipment to remain in the $l = 0$ age bracket for more than one period. For example, if the equipment is purchased in period $k = 0$, we may have $\mathbf{x}_{e,j}^{0,0}$, $\mathbf{x}_{e,j}^{1,0}$, and $\mathbf{x}_{e,j}^{2,0} = 1$ for the same e and j.

We solve this problem by counting a purchase, for a truck or loader of type e with identification j, as being made in period k if

$$\sum_l \mathbf{x}_{e,j}^{k,l} - \sum_l \mathbf{x}_{e,j}^{k-1,l} + \sum_l \mathbf{s}_{e,j}^{k,l}$$

is 1. The first term counts if we own the equipment in period k. The second term subtracts any equipment that we already owned. The third term prevents miscounting in the case where equipment was owned in the previous period but not owned in this period. Note that pre-existing equipment does not incur a capital expense as it was purchased at a time not considered in the optimisation period.

We denote the capital expense associated with equipment type e as F_e. However, we must discount future costs into the present by multiplying by a net present value discount factor. If I is the interest rate, this factor is

$$D_1^k = \frac{1}{(1+I)^k}. \tag{8.1}$$

Combining these terms, we see that the total capital expense for a truck or loader of type e with identification j is

$$F_e D_1^0 \mathbf{x}_{e,j}^{0,0} + \sum_{k>0} F_e D_1^k \left(\sum_l \mathbf{x}_{e,j}^{k,l} - \sum_l \mathbf{x}_{e,j}^{k-1,l} + \sum_l \mathbf{s}_{e,j}^{k,l} \right).$$

To represent the variable operating costs, we look at the utilisation variables to find the amount of time each piece of equipment is used. However, costs can vary from period to period, and also with how old a piece of equipment is. We denote the cost associated with running a truck or loader of type e in time period k and age bracket l by $V_e^{k,l}$. However, this must also be discounted to the net present value. Therefore, the total operating cost for an equipment of type e with identification j is

$$\sum_{k,l} V_e^{k,l} D_1^k \mathbf{f}_{e,j}^{k,l}.$$

Where salvage is advantageous, trucks and loaders can be retired at some benefit. As we are minimising cost rather than maximising profit, we represent salvage by a negative expense. To calculate the salvage value, we apply a depreciation rate,

J, to the capital expense of the equipment. We base depreciation on the age of the equipment l, rather than the time since it was bought. Combining depreciation with net present value discounting gives us the discount factor

$$D_2^{k,l} = \frac{(1-J)^l}{(1+I)^k}. \tag{8.2}$$

Since F_e is the original capital expense, the salvage 'cost' associated with equipment of type e and identification j is

$$-\sum_{k,l} F_e D_2^{k,l} \mathbf{s}_{e,j}^{k,l}.$$

Summing over all the equipment used, our complete objective function (which we wish to minimise) is:

$$\sum_{e,j} F_e D_1^0 \mathbf{x}_{e,j}^{0,0} + \sum_{e,j,k>0} F_e D_1^k \left(\sum_l \mathbf{x}_{e,j}^{k,l} - \sum_l \mathbf{x}_{e,j}^{k-1,l} + \sum_l \mathbf{s}_{e,j}^{k,l} \right) + \sum_{e,j,k,l} V_e^{k,l} D_1^k \mathbf{f}_{e,j}^{k,l} - \sum_{e,j,k,l} F_e D_2^{k,l} \mathbf{s}_{e,j}^{k,l}.$$

The first two terms represent the cost of capital outlay for the fleets; the next term represents the salvage cost of the fleet; and the final term denotes the operating expense.

8.2.4 *Constraints*

8.2.4.1 Basic Constraints

The necessary constraints for any equipment selection model are the *production constraints* and the *compatibility constraints*. The satisfaction of production requirements is the simplest constraint. In this model we need only account for the utilised production of the equipment rather than the potential production. The production capability, $P_e^{k,l}$, of a piece of equipment type e aged l in period k is determined by its availability (a_e^l), capacity (c_e) and cycle time (t_e^k):

$$P_e^{k,l} = \frac{a_e^l c_e}{t_e^k}. \tag{8.3}$$

Thus we have the following production constraints where T^k is the production requirement for period k:

$$\sum_{e,j,l} P_e^{k,l} \mathbf{f}_{e,j}^{k,l} \geq T^k \quad \forall \ k. \tag{8.4}$$

However, we need only generate these constraints for all loaders if we include the following balance equations:

$$\sum_{i,j,l} P_i^{k,l} \mathbf{f}_{i,j}^{k,l} = \sum_{i',j,l} P_{i'}^{k,l} \mathbf{f}_{i',j}^{k,l} \quad \forall \ k, i \in \mathbf{X}, i' \in \mathbf{X}'. \tag{8.5}$$

Determining a suitably compatible fleet is also important in terms of satisfying production requirements. To do this, we need to consider every possible combination of interacting equipment.

We let **A** represent the set of all possible combinations of loaders:

$$\sum_{i \in X(A'),j,l} P_i^{k,l} \mathbf{f}_{i,j}^{k,l} \geq \sum_{i' \in A',j,l} P_{i'}^{k,l} \mathbf{f}_{i',j}^{k,l} \quad \forall \ A' \in \mathbf{A}, k. \tag{8.6}$$

As a power set of equipment types this will generate $k(2^n - 1)$ constraints. For the 13 period, 27 loader problem this equates to 1744830451 constraints. We recognise that for a given case study the number of loaders possible in the final solution will be much lower than the complete set. In this case we can limit the generation of constraints to a maximum of α loader types. This will produce $k \sum_{a=1}^{\alpha} (\frac{n!}{a!(n-a)!})$ constraints. For the 13 period, 27 loader problem with a maximum of 4 loader types, this equates to just 271089 constraints.

We solve the model with only the first level compatibility constraint (where A' is just one loader type), and only add the remaining constraints into the model post hoc if they are violated by the solution before re-solving using a separation algorithm.

8.2.4.2 Variable Transition

Time periods and age brackets do not necessarily increase in tandem, so we know that whenever a time period is incremented, a piece of equipment may remain in the same age bracket, or it may graduate to the next age bracket. Note that even if it is utilised at full capacity, it cannot increase two age brackets. This is because of the chosen age-bracket size (5000 h) compared to the maximum utilised hours of the equipment (typically 3500). Taking into account the possibility of salvage, this corresponds to the constraints (where $L^k(e) = \min\{P(e) + k - 1, L(e)\}$ and $P(e)$ is the highest age of pre-existing equipment type e at the start of the schedule):

$$\mathbf{x}_{e,j}^{k,l} + \mathbf{s}_{e,j}^{k,l} \leq \mathbf{x}_{e,j}^{k-1,l} + \mathbf{x}_{e,j}^{k-1,l-1} \quad \forall k > 0, l \in [1, L^k(e) - 1], e, j \tag{8.7}$$

$$\mathbf{x}_{e,j}^{k,l} + \mathbf{s}_{e,j}^{k,l} \leq \mathbf{x}_{e,j}^{k-1,l-1} \quad \forall k > 0, l = L^k(e) - 1, e, j \tag{8.8}$$

$$\mathbf{x}_{e,j}^{k-1,l-1} \leq \mathbf{x}_{e,j}^{k,l-1} + \mathbf{s}_{e,j}^{k,l-1} + \mathbf{x}_{e,j}^{k,l} + \mathbf{s}_{e,j}^{k,l} \quad \forall k > 0, l \in [1, L^k(e) - 1], e, j. \tag{8.9}$$

Constraint (8.7) ensures that we do not own or salvage equipment in period k if we did not at least own the equipment in the previous period $k - 1$. Constraint (8.8) is an ammendment of constraint (8.7) for the variables that lie on the diagonal edge of the variable matrix. Constraint (8.9) ensures that if we own a piece of equipment in period $k-1$, the we must at least own the equipment in the next period, k, in either the same or next age-bracket, or we must salvage the equipment in the next period, k, in either the same or next age-bracket.

As noted before, we only set one $\mathbf{x}_{e,j}^{k,l}$ to be 1 for any particular piece of equipment and time period:

$$\sum_{l} \mathbf{x}_{e,j}^{k,l} + \mathbf{s}_{e,j}^{k,l} \leq 1 \quad \forall \; k, e, j. \tag{8.10}$$

8.2.4.3 Age Bracket Graduation

The index l is dependent on the accrued utilised hours of the equipment with identification j. For this model we represent the operating hours of the mine by O^k, and use age brackets of size B_0 hours.

Definition 8.1 The **accumulated utilised hours** in age brackets of a piece of equipment at the start of period k is given by

$$\sum_{h<k,l} \frac{\mathbf{f}_{e,j}^{h,l} O^h}{B_0},$$

where e is the equipment type and j is the individual equipment identification number.

Lemma 8.1 *If ε is an arbitrarily small continuous number and the equipment is in age bracket l at the start of period k (equivalent to $\mathbf{x}_{i,j}^{k,l} = 1$), then*

$$l \leq \sum_{h<k,l} \frac{\mathbf{f}_{e,j}^{h,l} O^h}{B_0} \leq l + 1 - \varepsilon.$$

Lemma 8.2 *If we represent the accrued utilised hours of equipment type e with identification j in period k and age bracket l by:*

$$\sum_{h<k,l} \frac{\mathbf{f}_{e,j}^{h,l} O^h}{B_0},$$

then

$$l \leq \sum_{h<k,l} \frac{\mathbf{f}_{e,j}^{h,l} O^h}{B_0} \leq l + 1 - \varepsilon$$

is true when $\mathbf{x}_{e,j}^{k,l} = 1$ if

$$\sum_{h<k,l} \frac{H^h}{B_0} \mathbf{f}_j^{h,l} \geq M(\mathbf{x}_j^{k,l} - 1) + l \quad \forall \; k, l, j, \tag{8.11}$$

and

$$\sum_{h<k,l} \frac{H^h}{B_0}\mathbf{f}_j^{h,l} \leq l + 1 - \varepsilon + M(1 - \mathbf{x}_j^{k,l}) \quad \forall \ k, l, j, \tag{8.12}$$

where M is a large integer and ε is an arbitrarily small continuous number.

Proof The accumulated utilised hours, $\sum_{h<k,l} \mathbf{f}_{e,j}^{h,l} O^h$, is equivalent to the equipment age in hours for period k. Thus $\sum_{h<k,l} \mathbf{f}_{e,j}^{h,l} O^h \geq 0$. Note that $\mathbf{x}_{e,j}^{k,l}$ is a binary variable.

If $\mathbf{x}_{e,j}^{k,l} = 0$, then (8.11) becomes

$$\begin{aligned} \sum_{h<k,l} \frac{\mathbf{f}_{e,j}^{h,l} O^h}{B_0} &\geq M\mathbf{x}_{e,j}^{k,l} + l - M \\ &= l - M \\ &\geq -M. \end{aligned} \tag{8.13}$$

The inequality (8.13) is true since $\sum_{h<k,l} \frac{\mathbf{f}_{e,j}^{h,l} O^h}{B_0} \geq 0$. (8.12) becomes

$$\begin{aligned} \sum_{h<k,l} \frac{\mathbf{f}_{e,j}^{h,l} O^h}{B_0} &\leq M + ((l+1) - M - \varepsilon)\mathbf{x}_{e,j}^{k,l} \\ &= M. \end{aligned} \tag{8.14}$$

The inequality (8.14) is true since M is an arbitrarily large integer.

If $\mathbf{x}_{e,j}^{k,l} = 1$, then (8.11) becomes

$$\begin{aligned} \sum_{h<k,l} \frac{\mathbf{f}_{e,j}^{h,l} O^h}{B_0} &\geq M\mathbf{x}_{e,j}^{k,l} + l - M \\ &= M + l - M \\ &= l. \end{aligned} \tag{8.15}$$

The inequality (8.15) is true from Lemma 8.1. (8.12) becomes

$$\begin{aligned} \sum_{h<k,l} \frac{\mathbf{f}_{e,j}^{h,l} O^h}{B_0} &\leq M + ((l+1) - M - \varepsilon)\mathbf{x}_{e,j}^{k,l} \\ &= M + l + 1 - M - \varepsilon \\ &= l + 1 - \varepsilon. \end{aligned} \tag{8.16}$$

The inquality (8.16) is also true from Lemma 8.1.

■

We note that M does not need to be arbitrarily large, but merely at least as big as the maximum number of age brackets. To reduce computation time, M can be set to the maximum number of age brackets.

Lemma 8.2 excludes pre-existing equipment because the cumulative utilisation is calculated slightly differently for this case. However, the age bracket constraints must also hold for pre-existing equipment. If we define $P(e, j)$ for $(e, j) \in \mathbf{P}$ to be the age in hours of the pre-existing equipment of type e and identification j, then at time period k the accumulated utilised hours of this piece of equipment is

$$\sum_{h<k,l} \mathbf{f}_{e,j}^{h,l} O^h + P(e, j).$$

Constraints (8.11) and (8.12) can then be repeated for $(e, j) \in \mathbf{P}$ by replacing the accumulated utilised hours by the above term.

8.2.4.4 Salvage Restriction

Each individual piece of equipment can only be salvaged once:

$$\sum_{k,l} \mathbf{s}_{e,j}^{k,l} \leq 1 \quad \forall \; e, j. \tag{8.17}$$

We also cannot salvage equipment which is not owned (in the previous period):

$$\mathbf{s}_{e,j}^{k,l} \leq \mathbf{x}_{e,j}^{k-1,l} + \mathbf{x}_{e,j}^{k-1,l-1} \quad \forall \; k > 0, l > 0, e, j. \tag{8.18}$$

We do not permit an identification number to be re-used once the equipment has been salvaged:

$$\sum_{h<k,l} \mathbf{s}_{e,j}^{h,l} + \sum_{l} \mathbf{x}_{e,j}^{k,l} \leq 1 \quad \forall \; k, e, j. \tag{8.19}$$

Finally, we give a value to the unbounded salvage variable:

$$\mathbf{s}_{e,j}^{0,0} = 0 \quad \forall \; \{e, j\} \notin \mathbf{P}. \tag{8.20}$$

8.2.4.5 Pre-existing Equipment

We set each pre-existing equipment to be either selected or salvaged in the first period:

$$\mathbf{x}_{e,j}^{0,l} + \mathbf{s}_{e,j}^{0,l} = 1 \quad \forall \; (e, j) \in \mathbf{P}, l = P(e, j). \tag{8.21}$$

8.2.4.6 Symmetry-Breaking Constraints

In creating many variables that are identical save for the identification number, j, we have created large clusters of solutions that are effectively permutations of each other. This is not ideal computationally. However, we can eliminate this redundancy if we define some simple rules regarding which identification numbers to use first. We nominate that we would like to keep the equipment with the lowest identification the longest, bringing us to our final constraint for this model:

$$\mathbf{x}_{e,j}^{k,l} \geq \mathbf{x}_{e,j+1}^{k,l} \qquad \forall \ (e, j) \notin \mathbf{P}, k, l.$$

8.2.5 Complete Model

$$\text{Minimise} \sum_{e,j} F_e D_1^0 \mathbf{x}_{e,j}^{0,0} + \sum_{e,j} \left(\sum_{k>0,i} F_e D_1^k \Big(\sum_l \mathbf{x}_{e,j}^{k,l} - \sum_l \mathbf{x}_e^{k-1,l} + \sum_l \mathbf{s}_{e,j}^{k,l}\Big) \right)$$
$$+ \sum_{k,e,j,l} V_e^l D_1^k H^k \mathbf{f}_e^{k,l} - \sum_{k,e,j,l} F_e D_2^{k,l} \mathbf{s}_{e,j}^{k,l}$$

$$\text{s.t.} \quad \sum_{i',l} P_{i'}^{k,l} \mathbf{f}_{i'}^{k,l} \geq T^k \qquad \forall\, k, \tag{8.22}$$

$$\sum_{i<m+1,l} P_i^{k,l} \mathbf{f}_i^{k,l} = \sum_{i'>m,l} P_{i'}^{k,l} \mathbf{f}_{i'}^{k,l} \qquad \forall\, k, \tag{8.23}$$

$$\sum_{i \in X(A'), i<m+1,l} P_i^{k,l} \mathbf{f}_i^{k,l} \geq \sum_{i' \in A', i'>m,l} P_{i'}^{k,l} \mathbf{f}_{i'}^{k,l} \qquad \forall\, A' \in \mathbf{A}, k, \tag{8.24}$$

$$\mathbf{x}_i^{k,l} + \mathbf{s}_{e,j}^{k,l} \leq \mathbf{x}_i^{k-1,l} + \mathbf{x}_i^{k-1,l-1} \qquad \forall\, k > 0, l \in [1, L^k(i) - 1], i, \tag{8.25}$$

$$\mathbf{x}_i^{k,l} + \mathbf{s}_{e,j}^{k,l} \leq \mathbf{x}_i^{k-1,l-1} \qquad \forall\, k > 0, l = L^k(i) - 1, i, \tag{8.26}$$

$$\mathbf{x}_i^{k-1,l-1} \leq \mathbf{x}_i^{k,l-1} + \mathbf{s}_i^{k,l-1} + \mathbf{x}_{e,j}^{k,l} + \mathbf{s}_{e,j}^{k,l} \qquad \forall\, k > 0, l \in [1, L^k(i) - 1], i, \tag{8.27}$$

$$\sum_l \mathbf{x}_i^{k,l} \leq 1 \qquad \forall\, k, i, \tag{8.28}$$

$$\sum_{k'<k,l'<l} \frac{\mathbf{f}_i^{k',l'} H^{k'}}{B_0} \geq M(\mathbf{x}_i^{k,l} - 1) + l \qquad \forall\, k > 0, i, l, \tag{8.29}$$

$$\sum_{k'<k,l'<l} \frac{\mathbf{f}_i^{k',l'} H^{k'}}{B_0} \leq l + 1 - \varepsilon + M(1 - x) \qquad \forall\, k > 0, i, l, \tag{8.30}$$

$$\mathbf{f}_i^{k,l} \leq \mathbf{x}_i^{k,l} \qquad \forall\, k, i, l, \tag{8.31}$$

$$\mathbf{s}_i^{k,l} \leq \mathbf{x}_i^{k-1,l} + \mathbf{x}_i^{k-1,l-1} \qquad \forall\, k > 0, i, l > 0, \tag{8.32}$$

$$\sum_{k'<k,l} \mathbf{s}_i^{k',l} + \mathbf{x}_{e,j}^{k,l} \leq 1 \qquad \forall\, k, i, l, \tag{8.33}$$

$$\mathbf{s}_{e,j}^{0,0} = 0 \qquad \forall\, \{e, j\} \notin \mathbf{P} \tag{8.34}$$

$$\mathbf{x}_{e,j}^{0,l} + \mathbf{s}_{e,j}^{0,l} = 1 \qquad \forall\, (e, j) \in \mathbf{P}, e, l \tag{8.35}$$

$$\mathbf{x}_i^{k,l} \geq \mathbf{x}_{i+1}^{k,l} \qquad \forall\, i \notin \mathbf{P}, k, l \tag{8.36}$$

$$\mathbf{x}, \mathbf{s} \in \{0, 1\}$$

$$\mathbf{f}, \in [0, 1]$$

8.3 Validation Test Case

There are two modes of model validation that we analyse here. In the first instance, we ensure that the model behaves as we expect. Secondly, we show that the model adequately captures the real world problem.

To ensure that the model behaves as we expect, we created a test case that show that the following actually occurs:

- Purchase and salvage can occur in any time period;
- The production requirements are met;
- The fleets are sufficiently compatible (subject to production requirements);
- The age index, l, is actually capturing the cumulative utilisation;
- Equipment does not age more than one bracket at a time;
- Equipment is salvaged at most once and only if it was owned in the previous period;
- Equipment is not used beyond its retirement age;
- Pre-existing equipment is either kept or immediately salvaged.

Therefore we need the test case to run for such a length of time that some equipment reach the end of their replacement cycle. We must also include pre-existing equipment, and start from a set of trucks and loaders that are not all compatible. To do this, we created a test case with two types of trucks and two types of loaders. We introduced four pre-existing trucks of type 0, one pre-existing loader of type 0, and two pre-existing loaders of type 1. We set up simple mine schedule characteristics such as constant production requirements at a relatively low rate of 4 million tonnes per period; and a constant truck cycle time of 10.2 min. Also, we simplified

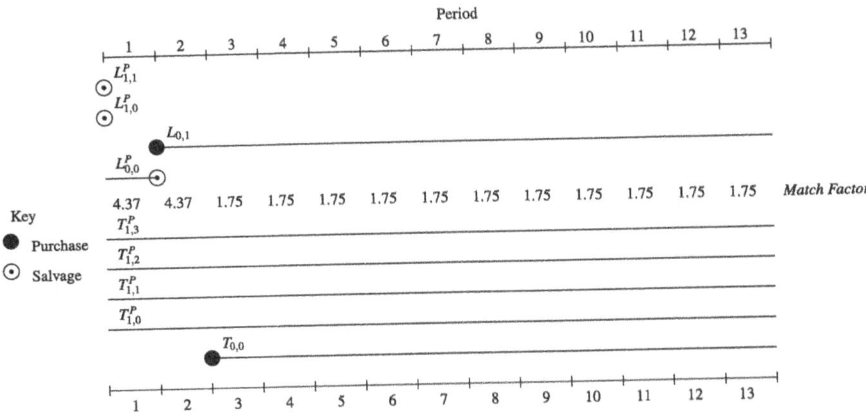

Fig. 8.3 The optimal purchase and salvage policy for the 13-period validation test case

the remaining parameters in the model, such as constant operating expense over the entire schedule, and unvarying availability of equipment. Lastly, we reduce the maximum life of the trucks to 35000 h to capture the replacement cycle of the equipment in the solution.

We programmed the model in *C++* using *Ilog Concert Technology v2.5* objects, and solved the program with default MILP algorithms in *Ilog Cplex v11.0*. We implemented the validation problem with 14496 variables and 24280 constraints. We included only two levels of the compatibility power constraints (8.6). This is possible because the optimal solution contains only two loader types, so a higher level is redundant. The problem solved using the default Cplex parameters in 1893 s (32 min). In the solution, one new loader was purchased at the beginning of period 2 and one new truck was purchased at the beginning of period 3 (Fig. 8.3). All pre-existing trucks are kept for the entire schedule; the two pre-existing loaders (type 1) are salvaged at the beginning of the schedule while the loader of type 0 is retained for only one period.

In Table 8.1 we sum the utilised hours of the new truck to represent the age of the equipment over the accummulating periods. By dividing by the size of the bracket (5000), we calculate the age of the truck in terms of age brackets. By comparing this value to the solution value, we can determine whether the implemented model is behaving as we expect. We can see that these values match, which indicates that the age bracket index, l, is effectively capturing the cumulative utilisation of the equipment. We can also see that, as we expect, the equipment never ages more than one bracket at a time. Thus we are satisfied that we have implemented the model correctly.

Table 8.1 The validation of the matching of ageing index with actual cumulative age

	Period												
$T_{0,0}$	1	2	3	4	5	6	7	8	9	10	11	12	13
Actual age (hours)	0	0	2299.46	4598.92	6898.37	9197.83	11497.3	13796.7	16096.2	18395.7	20695.1	22994.6	25294
Actual age (brackets)	0	0	0	0	1	1	2	2	3	3	4	4	5
Solution age (brackets)	0	0	0	0	1	1	2	2	3	3	4	4	5

8.4 Computational Study

We implemented this model on the *many locations* case study presented in Chap. 5. We start by analysing the first four periods of this thirteen period case study. The 4-period model contained 65940 variables and 119656 constraints before the higher level compatibility constraints were added. After performing preliminary tests, we chose a depreciation value of 53% which obtained the fastest solution time. The *Cplex* optimiser found the optimal solution in 1582 s (27 min) with an overall fleet cost of $\$5.34279e^7$. We analyse this solution first. After the initial solve, no higher level compatibility constraints were found to be breached and the solution was accepted as optimal. The solution converged to within 3% of optimality after just one minute of computation, taking the remaining 26 min to close this small gap (Fig. 8.4).

In the optimal solution, all the pre-existing loaders were salvaged: reflecting both their age and their low compatibility with other "cheap" trucks (Fig. 8.5). Two new loaders were purchased: one loader of type 3 in period 1, and one loader of type 25 in period 2. The presence of only two loader types in the solution calls a need for up to only two levels of compatibility constraint. In the truck fleet, three types were selected overall: the type-12 pre-existing trucks were kept for the entire four periods; one of the medium-sized type-8 trucks was purchased in period-1 and nine of the super-sized type-9 trucks was purchased in period-2. This reflects both the increasing demands of the production schedule and increased truck cycle times, and the declining performance of the older pre-existing trucks.

One note-worthy aspect of this solution is that there are still three truck types being selected. As an optimal solution, this goes against the intuition of the mining engineer, who has a greater sense of the hidden costs of operating multiple types of equipment. Unfortunately these costs were not visible to the model, and this is reflected in the types of solutions it finds. This clearly reveals a need for better cost estimates for heterogeneous fleets.

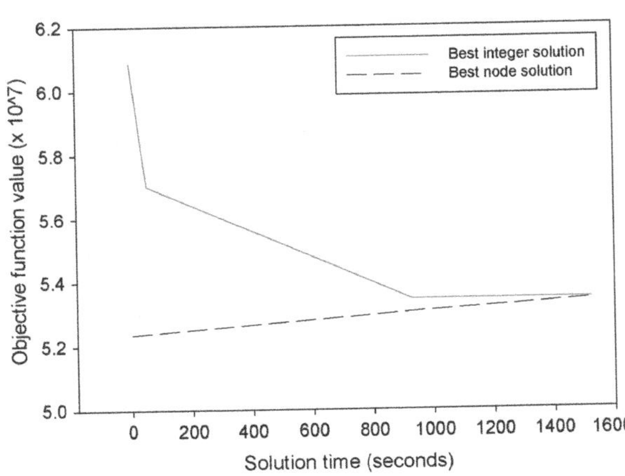

Fig. 8.4 Convergence of the 4-period utilisation cost model, with the depreciation parameter set to 53%

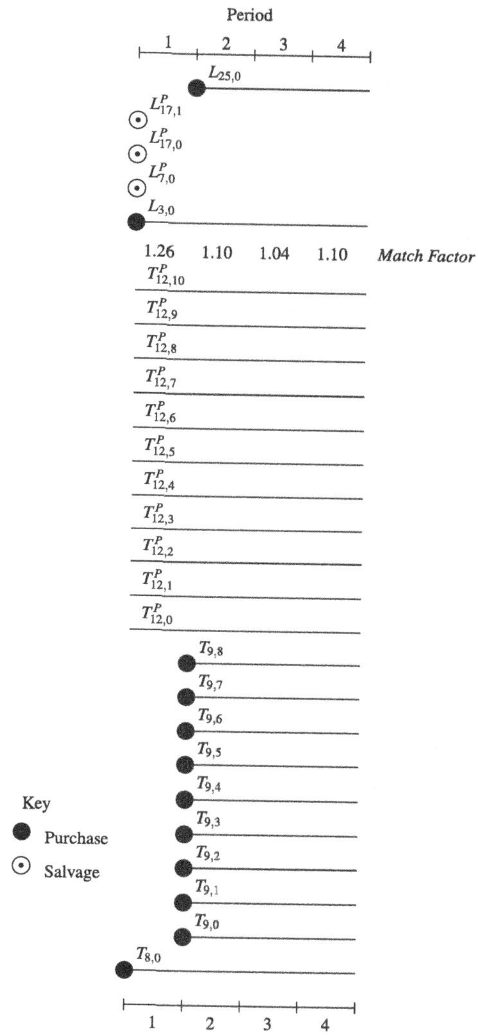

Fig. 8.5 The optimal solution for the 4-period case study with the utilisation model. The depreciation parameter is set to 53%

The match factor for each period reveals that the fleet is well balanced, and that the truck fleet has some slackness. Again, this reflects the greater cost of running the loaders over the trucks.

An important part of the way we modelled this problem is the optimisation of the utilisation of the equipment (Table 8.2). That is, as we are accounting for the way that the equipment is used in each period, we are able to optimise the allocation of equipment to tasks. This means that in the solution we also obtain an optimised utilisation policy. In this policy it is clear to see which pieces of equipment are not fully utilised. This provides the mining manager with some flexibility, and a clear indication of where these flexibilities occur in the fleet.

Table 8.2 The optimal utilisation policy for the 4-period case study, when the depreciation parameter is set to 53%

Equipment	Period			
	1	2	3	4
$T_{8,0}$	1	1	1	1
$T_{9,0}$	0	1	1	1
$T_{9,1}$	0	1	1	1
$T_{9,2}$	0	1	1	1
$T_{9,3}$	0	1	1	1
$T_{9,4}$	0	1	1	1
$T_{9,5}$	0	1	1	1
$T_{9,6}$	0	1	1	1
$T_{9,7}$	0	1	1	1
$T_{9,8}$	0	1	1	1
$T_{12,0}$	0.935386	0.4821	0	1
$T_{12,1}$	1	0	1	1
$T_{12,2}$	1	0	1	1
$T_{12,3}$	1	0	1	1
$T_{12,4}$	1	0	1	1
$T_{12,5}$	1	1	0	0.919593
$T_{12,6}$	1	1	0.232463	1
$T_{12,7}$	1	0.0112244	1	1
$T_{12,8}$	1	1	1	1
$T_{12,9}$	1	0.0112244	1	1
$T_{12,10}$	1	1	1	1
$L_{3,0}$	0.852581	1	1	1
$L_{25,0}$	0	0.0772999	0.709341	0.89751

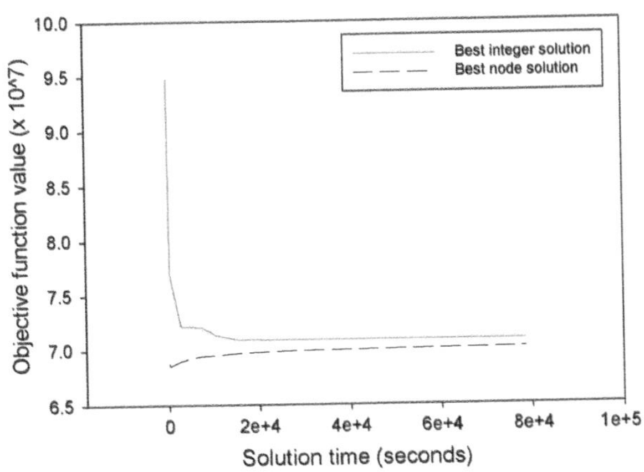

Fig. 8.6 Convergence of the 5-period utilisation cost model, with the depreciation parameter set to 53%

We implemented the 5-period problem with 95550 variables and 174037 constraints. After 70926 s (19.7 h) of computation we obtained a solution with optimality gap of 1.10%, with an objective function value of \$7.09709e^7. In terms of a mining operation where some aspects of the schedule are expected to change over time, this is not an unreasonable gap. We observe that the final solution (with optimality

Fig. 8.7 The solution for the 5-period case study with optimality gap 1.10% for the utilisation model. The depreciation parameter is set to 53%

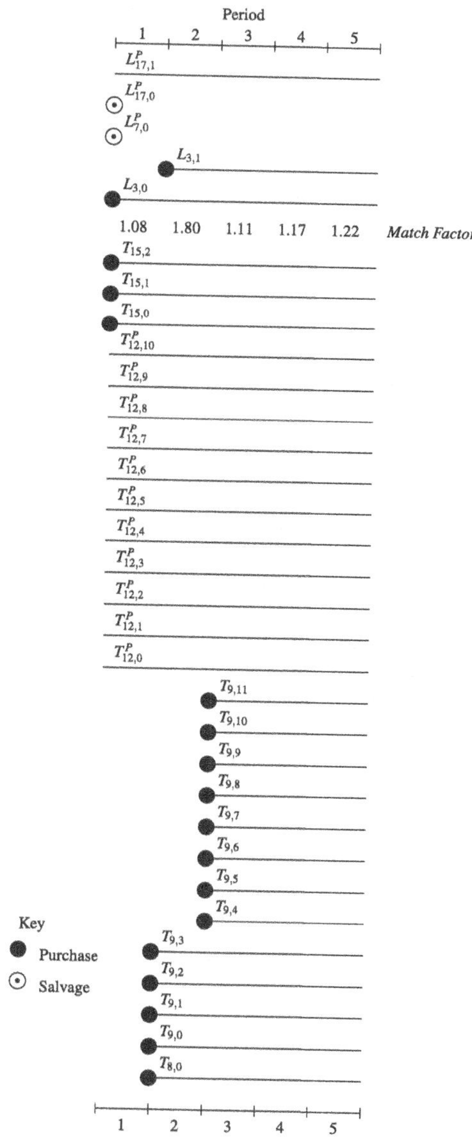

gap 1.10%) was found after 17647 s, although it took a further 60000 s to close the remaining 0.4% (Fig. 8.6).

It is interesting to observe the differences between the 4-period solution and the 5-period solution (Fig. 8.7). Instead of salvaging all the pre-existing loaders, we opt to keep the loader of type 17 and purchase two new type-2 loaders over the first two periods. This is related to the choice to purchase a fleet of type-15 trucks in the first period while keeping all the pre-existing trucks of type 12—in the 4-period optimal solution, no type-15 trucks were purchased at all. This highlights the need to optimise over the entire schedule, as the solutions for shorter periods can be quite different and lead to worse long-term solutions.

8.5 Sensitivity Analysis

The parameters in this model are subject to uncertainty in two ways. Some parameters are known to be estimates, such as the depreciation of equipment; some parameters are certain to change once the schedule begins and new information comes to light, such as truck cycle time and production requirements. In our sensitivity analysis, we are interested in the influence of these uncertain parameters on the robustness of the model. We begin by studying the behaviour of the model at differing values of depreciation (Table 8.3).

We observe an interesting phenomenon here where there exists a critical value near 53% where the solution time drops drastically (Fig. 8.8). However, the purchase and salvage policies are identical across all depreciation values tested. Instead of differing fleets, this critical point marks the decision to significantly alter the utilisation of several pre-existing trucks (Table 8.4). Interestingly, the objective function values for the 4-period case study converge as the depreciation value increases (Fig. 8.9).

Table 8.3 The 4-period case study solutions with varying depreciation

Periods	Depreciation(%)	Time (s)	Quality	Objective function value($)
4	35	2444	Optimal	5.33485×10^7
4	40	2608	Optimal	5.33968×10^7
4	45	2457	Optimal	5.34175×10^7
4	50	2717	Optimal	5.34258×10^7
4	51	2721	Optimal	5.34266×10^7
4	52	2274	Optimal	5.34273×10^7
4	53	1588	Optimal	5.34279×10^7
4	54	1764	Optimal	5.34284×10^7
4	55	1985	Optimal	5.34287×10^7
4	60	2386	Optimal	5.34297×10^7

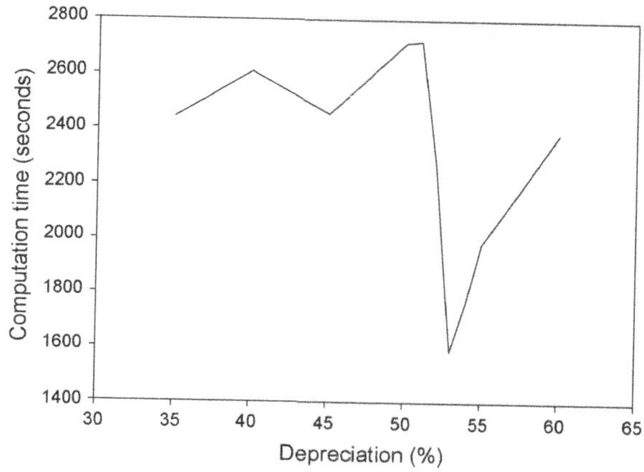

Fig. 8.8 The varying solution times with depreciation for the 4-period case study

Table 8.4 The utilisation policy for the 4-period case study with 53% depreciation

Equipment	Period			
	1	2	3	4
$T_{8,0}$	1	1	1	1
$T_{9,0}$	0	1	1	1
$T_{9,1}$	0	1	1	1
$T_{9,2}$	0	1	1	1
$T_{9,3}$	0	1	1	1
$T_{9,4}$	0	1	1	1
$T_{9,5}$	0	1	1	1
$T_{9,6}$	0	1	1	1
$T_{9,7}$	0	1	1	1
$T_{9,8}$	0	1	1	1
$T_{12,0}$	1	0	1	0.919593
$T_{12,1}$	1	0	0.210014	1
$T_{12,2}$	1	0	1	1
$T_{12,3}$	0.935386	0.4821	0	1
$T_{12,4}$	1	0	1	1
$T_{12,5}$	1	0.0112244	1	1
$T_{12,6}$	1	1	1	1
$T_{12,7}$	1	0.0112244	1	1
$T_{12,8}$	1	1	1	1
$T_{12,9}$	1	1	0.0224488	1
$T_{12,10}$	1	1	1	1
$L_{3,0}$	0.852581	1	1	1
$L_{25,0}$	0	0.0772999	0.709341	0.89751

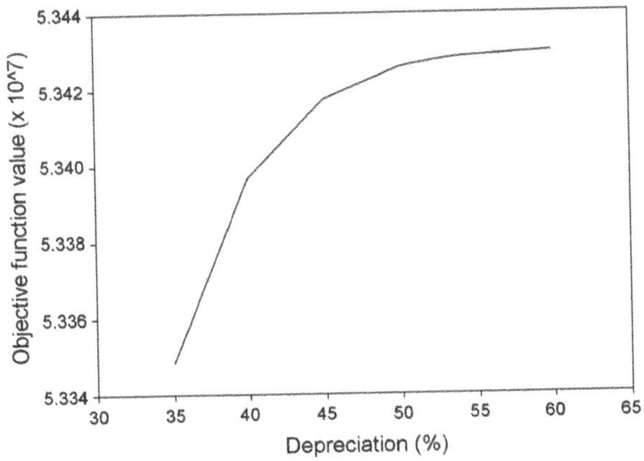

Fig. 8.9 The varying objective functions with depreciation for the 4-period case study

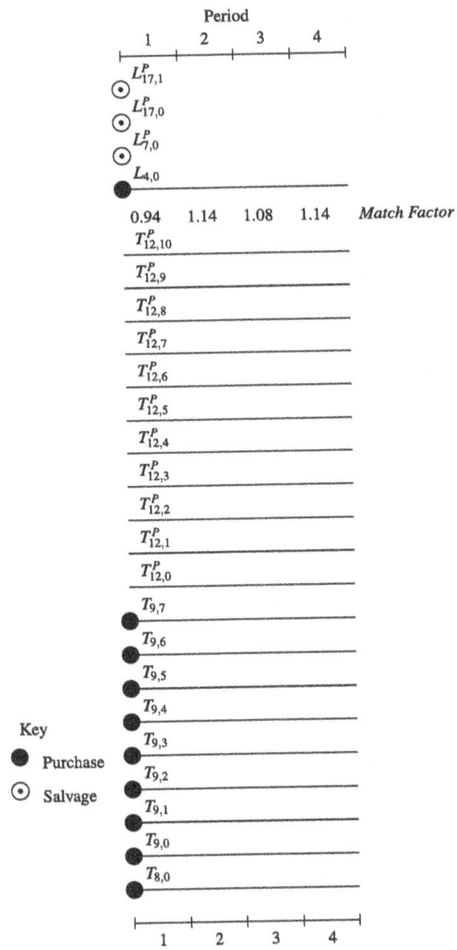

Fig. 8.10 We flatten the production requirements for the 4-period case study to observe the fleet changes due to ageing and truck cycle time

Next we study the influence of truck cycle time on the solution when the production requirements are uniform over the entire schedule to 36 million tonnes per period. Thus we expect any variability in the fleet to be exclusive to two factors: ageing of the equipment and truck cycle time. In a 4-period solution, the effects of ageing are not apparent (Fig. 8.10). This test case solved in 1576 s, demonstrating that a simplified production schedule does not necessarily simplify the problem. The decreasing slackness in the utilisation policy hints at the increasing truck cycle time, but otherwise the fleet is fairly stable (Table 8.5). In fact, with less volatility in the schedule, the purchase policy is also more stable than the case study solution—only one loader type is used.

We repeat this experiment for production requirements by flattening the truck cycle time over the entire schedule and observing the behaviour of the solution. We set the truck cycle times over the four periods to 30.75 min. In this case, we expect the purchasing behaviour to closely reflect the demands from the production requirements.

Table 8.5 The utilisation policy for the 4-period problem with flattened production requirements

Equipment	Period			
	1	2	3	4
$T_{8,0}$	1	1	1	1
$T_{9,0}$	1	1	1	0.829834
$T_{9,1}$	1	1	1	1
$T_{9,2}$	1	1	1	1
$T_{9,3}$	1	1	1	1
$T_{9,4}$	1	1	1	1
$T_{9,5}$	1	1	1	1
$T_{9,6}$	1	1	1	1
$T_{9,7}$	1	1	1	1
$T_{12,0}$	0.0224488	1	1	1
$T_{12,1}$	0.0224488	1	1	1
$T_{12,2}$	0.705255	1	0.317194	1
$T_{12,3}$	0.0224488	1	1	1
$T_{12,4}$	0.179235	0.843213	1	1
$T_{12,5}$	1	1	0.252479	1
$T_{12,6}$	1	1	1	1
$T_{12,7}$	1	1	1	1
$T_{12,8}$	1	1	1	1
$T_{12,9}$	1	1	1	1
$T_{12,10}$	1	1	1	1
$L_{4,0}$	0.998269	0.998269	0.998269	0.998269

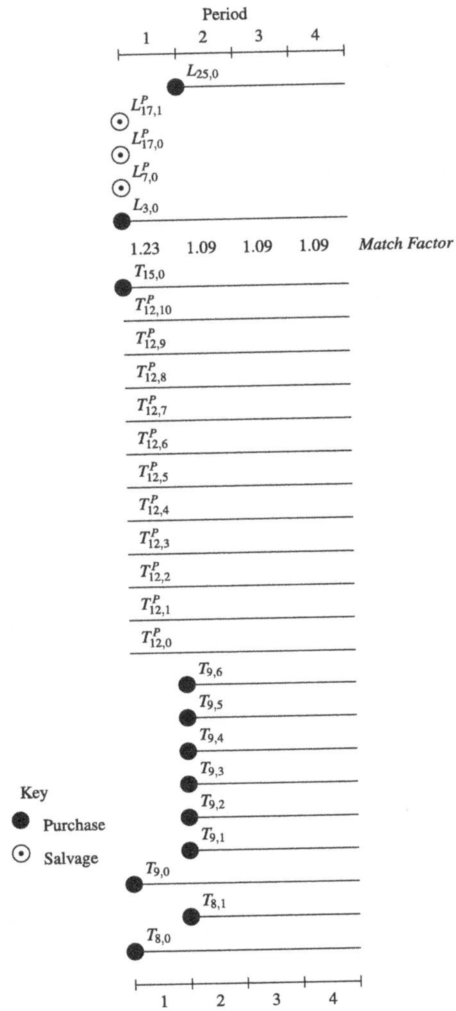

Fig. 8.11 We flatten the truck cycle times in the 4-period case study

We obtained the optimal solution to this test case in 5781 s (97 min) (Fig. 8.11). This indicates the increased complexity of the problem when the production requirements vary from period to period—there are many options to increase the size of the fleet to satisfy a small change in productivity, and many of these options still will be heterogeneous fleets. This is further confirmed by observing the level of heterogeneity in the fleet: two loader types and four truck types. The changes in the utilisation policy can be clearly noted to follow the changes in the production schedule (Table 8.6). That is, as the actual production requirements increase, so too does the utilisation of loaders until the fourth period where almost all equipment is used fully.

Table 8.6 The utilisation policy for the 4-period test case with flattened truck cycle times

Equipment	Period			
	1	2	3	4
$T_{8,0}$	1	0.0112244	1	1
$T_{8,1}$	0	1	1	1
$T_{9,0}$	1	0.0112244	1	1
$T_{9,1}$	0	1	1	1
$T_{9,2}$	0	1	1	1
$T_{9,3}$	0	1	1	1
$T_{9,4}$	0	1	1	1
$T_{9,5}$	0	1	1	1
$T_{9,6}$	0	1	1	1
$T_{12,0}$	0.697495	0	1	0.90094
$T_{12,1}$	1	0.98417	0.0382786	1
$T_{12,2}$	1	0	1	1
$T_{12,3}$	1	0	1	1
$T_{12,4}$	1	0	1	1
$T_{12,5}$	1	1	1	1
$T_{12,6}$	1	1	1	1
$T_{12,7}$	1	1	1	1
$T_{12,8}$	1	1	1	1
$T_{12,9}$	1	1	0.427323	1
$T_{12,10}$	1	1	1	1
$T_{15,0}$	1	1	1	1
$L_{3,0}$	0.852581	1	1	1
$L_{25,0}$	0	0.0772999	0.709341	0.89751

Other parameters that have great influence on the scale of the problem include:

- number of available truck types
- number of available loader types
- number of periods
- number of identification numbers per equipment type
- ε (the chosen tolerance value)
- number of age brackets (which corresponds to the Big-M value in the model)

Clearly, reducing the overall number of trucks, loaders and periods will improve the tractability and computation time of the model. As we observed with the 5-period problem, increasing the scale of the problem by increasing the number of periods, trucks or loaders only renders the problem too large to solve. We can reduce the number of identification numbers per equipment type to the optimal number and gain significant reductions in computation time. However, to do this we need to know the optimal number of identifications *a priori*, and this information is only gleamed after solving the problem. Before solving the problem, we must be cautious that we do not over-constrain the problem by setting the number of identifications at too low a level.

We expect the value of ε to have an impact on computation time (with a noticeable effect on solution only with large values of ε). We also expect that there is a critical value of ε that relates to the stability of the solution—as it is used in the age-bracket constraint, ε is scaled by the size of the age bracket (in this case study, 5000). We are therefore interested in studying the behaviour of the model as ε decreases in size. We do this by observing the changes in computation time *and* objective function value as we reduce ε. Figure 8.12 depicts the predicted unstable behaviour, before the solution stabilises at $\varepsilon = 0.000000001$ (Table 8.7).

The Big-M value needs to only be large enough to guarantee the validity of the age-bracketing constraints. Therefore, the number of age brackets corresponds to the Big-M value in the model. While this number is fixed by the policies of the mine, we are interested in studying the robustness of the model subject to different age bracket values. To do this we vary M size in increments of 5. We expect the computation time to increase with the size of age bracket, reflecting the increased difficulty in solving the linear program associated with the Big-M value. Again we study the problem with depreciation set to 53% and ε set to 0.00000001. We vary the size of M in four subsequent tests from 15 to 30 (Table 8.8).

We can see through the analyses of these uncertain parameters that the model is reasonably robust—the solutions themselves are stable and the main aspect affected by the uncertainty is computation time.

Other parameters subject to less variability (and are therefore not considered as interesting for our analysis) are:

- NPV discount rate
- operating hours of the mine
- compatibility of equipment
- availability of equipment
- fixed cost of equipment
- equipment capacity
- operating cost
- maximum age of equipment

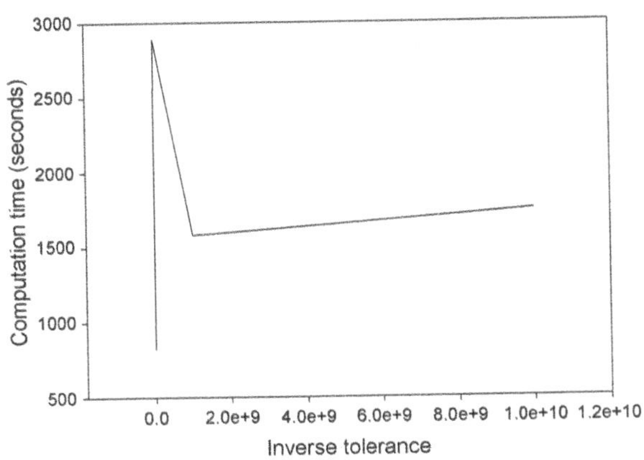

Fig. 8.12 The varying solution times as we vary the tolerance values (note that the tolerance values are inverted here)

Table 8.7 The variance of computation times and objective function value when we vary ε for the 4-period case study solutions with depreciation at 53%

Periods	ε	Time (s)	Quality	Solution
4	1e-10	1748	Optimal	5.34279×10^7
4	1e-9	1579	Optimal	5.34279×10^7
4	1e-8	1538	Optimal	5.34279×10^7
4	1e-7	2889	Optimal	5.34279×10^7
4	1e-6	2232	Optimal	5.34279×10^7
4	1e-5	2314	Optimal	5.34279×10^7
4	1e-4	824	Optimal	5.34279×10^7
4	1e-3	2080	Optimal	5.3428×10^7

Table 8.8 The variance of computation times and objective function value when we vary ε for the 4-period case study solutions with depreciation at 53%

Periods	Cost-brackets	Time (s)	Quality	Solution
4	15	10801	Optimal	5.12685×10^7
4	20	11154	Optimal	5.12685×10^7
4	25	13065	Optimal	5.12685×10^7
4	30	13065	Optimal	5.12685×10^7
4	40	5057	Optimal	5.12685×10^7

8.6 Conclusion

In this chapter, we have observed some interesting results. Initially we expected that by accounting for utilisation in the objective function we would eliminate the tendency to purchase multiple truck types. However, as we saw with the case study results, this is not the case. This suggests that high volatility in production requirements is best dealt with high heterogeneity—a fleet with mixed sizes is more flexible to changes in demands than a fleet of all the same size. This makes sense analogously if we consider that two 50 cent pieces cannot be used to closely match prices that are higher than 50 cents and less than 1 dollar in the same way that one 50 cent, two 20 cent and one 10 cent coins can.

As the equipment operating cost is now calculated on hours used, it is no longer necessary for owned equipment to operate in every period as before. This reduces the need to purchase equipment for short periods of time and then salvage it, resulting in a more accurate model than the single-location model presented in Chap. 6.

An interesting observation was the critical phenomena for the depreciation parameter—corresponding with a shift in utilisation policy. This result, along with other sensitivity analysis, demonstrates that the model is robust to uncertainty, and that the main factor affected by uncertainty is computation time—a pleasing result.

It is clear that the problem of equipment selection for surface mines is not trivial due to the dependencies of costs on the age of the equipment, which may be non-uniform. We had some success in improving the solvability of our model by introducing symmetry-breaking constraints and by with-holding higher-level compatibility constraints as lazy constraints. However, for this model to be useful in practice we would like to see greater scalability and the potential to consider large mining schedules (with up to, say, 20 periods). Greater scalability would enable us to consider a longer number of mining and dumping locations: allowing us to solve the model in a more realistic setting.

In spite of this shortcoming, we are satisfied that we adequately captured the non-uniform piece-wise linear ageing of the equipment within linear constraints. Further, we have presented a model that considers pre-existing equipment and heterogeneous fleets, allows equipment to be salvaged, and outputs a utilisation policy, and a purchase and salvage policy. This model is robust to the uncertainty of a mining schedule, and provides the mining manager with a set of optimal decisions based on the best available information at the time of solution.

Notation Index

A summary of the notation used in this chapter is as follows.

- $\mathbf{X}$ the set of all available new truck types.
- $\mathbf{X}'$ the set of all available new loader types.
- i the truck type index, $i \in \mathbf{X}$.
- i' the loader type index, $i' \in \mathbf{X}'$.
- e the truck or loader type index, e, where $[e \in \{i, i'\} | i \in \mathbf{X}, i' \in \mathbf{X}']$.
- j the individual machine identification index.
- k the period index.
- l the equipment age bracket index.
- m the number of available new truck types.
- n the number of available new loader types.
- M an arbitrarily large integer.
- ε an arbitrarily small continuous number.
- $\mathbf{P}$ the set of all pre-existing equipment types.
- $\mathbf{X}(i')$ the set of all truck types compatible with loader $i' \in \mathbf{Y}$.
- K the total number of periods in the mining schedule.
- $L(e)$ the maximum age (in operating hours) unique for equipment type e.
- $L^k(e)$ the minimum of the current period size plus pre-existing age, and maximum age, $L^k(e) = \min\{P(e) + k, L(e)\}$.
- $\mathbf{x}_{e,j}^{k,l}$ indicator variable for truck or loader type e with machine id j selected in period k, while in age bracket l.
- $\mathbf{f}_{e,j}^{k,l}$ the proportion of time that truck type e with machine id j is utilised in period k, in age bracket l.
- $\mathbf{s}_{e,j}^{k,l}$ the number of trucks type e salvaged in period k, aged l.
- $\mathbf{g}_{i'}^{k}$ the indicator variable denoting selection of loader type $i' \in \mathbf{X}'$ in period k.
- $\mathbf{h}_{i'}^{k}$ the indicator variable denoting constraint dominance for loader type $i' \in \mathbf{X}'$ in period k.

$x_e^{P(e,j)}$ indicator variable for pre-existing equipment type $e \in \mathbf{P}$ with id j at the start of the schedule.
F_e the fixed cost (capital expense) of obtaining equipment type e.
D_1^k the discount factor for period k.
$D_2^{k,l}$ the discount factor and depreciation factor for period k for equipment aged l.
$V_e^{k,l}$ the variable expense for equipment type e aged l in period k.
$P_e^{k,l}$ the productivity of equipment e in period k at age l.
T^k the required productivity of the mine (in tonnes) for period k.

Acknowledgements Components of this chapter were presented in [1].

Reference

1. C. Burt, An optimisation approach to materials handling in surface mines. Ph.D. thesis, Mathematics and Statistics, Curtin University of Technology, Bentley, Perth, Australia, 2008, http://espace.library.curtin.edu.au:80/R?func=dbin-jump-full&local_base=gen01-era02&object_id=165737

Chapter 9
Accurate Costing of Mining Equipment

Christina N. Burt and Yao-ban Chan

9.1 Introduction

In equipment selection, it is known that operating expense dominates the cost of materials handling over time [2]. The operating expense reflects the cost of operating and maintaining the equipment. It takes into account varying maintenance expenses, availability and productivity levels, all of which are known to vary with the age of the equipment. We can best account for the operating cost by considering the number of hours that the equipment operates; equipment is often not utilised to full capacity, and not accounting for this difference may lead to inferior solutions. However, including cost as a function of utilisation and equipment age adds great complexity to the problem, and has the potential to introduce nonlinearities [3].

In the mining industry, the nonlinear operating cost is commonly simplified by creating a piecewise linear function that is divided into age brackets, so that the cost is constant while the equipment is in any given age bracket (Fig. 9.1). While this discretisation allows us to easily apply linear techniques (such as integer or mixed-integer programming) to model problems involving the costs, it has disadvantages when it comes to actually calculating this cost. Traditionally, it is assumed that a unit

---

The original version of this chapter was revised: Originator type has been changed and the chapter author names have been included. The erratum to this chapter is available at https://doi.org/10.1007/978-3-319-76255-5_11

C. N. Burt (✉)
Department of Mathematics and Statistics, The University of Melbourne,
Parkville, VIC, Australia
e-mail: c.burt@ms.unimelb.edu.au

Y.-b. Chan
The University of Melbourne, Melbourne, VIC, Australia
e-mail: yaoban@unimelb.edu.au

C. N. Burt and L. Caccetta (eds.), *Equipment Selection for Mining: With Case Studies*, Studies in Systems, Decision and Control 150,
https://doi.org/10.1007/978-3-319-76255-5_9

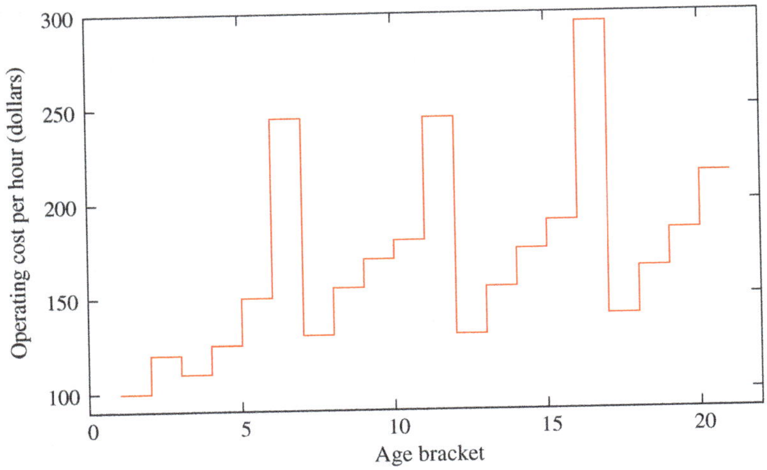

Fig. 9.1 The variability of equipment operating cost against age bracket. The rise in operating cost is due to increased maintenance expense over time. Large drops in the operating expense occur when an overhaul has taken place

stays in the same age bracket for an entire period of time, and so the cost is calculated based on the age of the equipment at the start of that period.

Unfortunately, this is an unrealistic assumption. In many cases, a piece of equipment will sit in the starting age bracket for only a fraction of the period, and spend the remainder of the period in the next bracket, resulting in a different cost to the one calculated. Clearly this will introduce error into the objective function. On the other hand, since time periods are generally shorter than age brackets, it is also possible that the equipment may stay in the same age bracket for the entire period. These two possibilities are illustrated in Fig. 9.2. To accurately account for cost in case (b), we have to determine the amount of time that the unit spends in the next age bracket.

In Chap. 8, we presented a model for mining equipment selection accounting for utilisation of equipment. In this chapter, we extend these results by presenting a method for determining the actual cost when equipment moves into the next age bracket during a period. Section 9.2 shows how accurate costing can be achieved in a

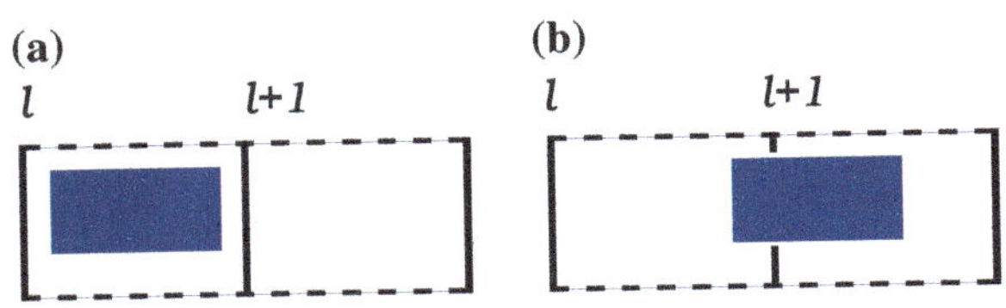

Fig. 9.2 Possible scenarios that occur within a time period. In case (**a**), the unit stays in the same age bracket for the entire period. In case (**b**), the unit moves into the next age bracket within the period

non-utilisation model. In Sect. 9.3, we observe how we can account for utilisation and age brackets in an equipment selection model. In Sect. 9.4, we show how accurate costing can also be achieved for a utilisation model. Finally, we note in Sect. 9.5 how we can use this framework to improve the accuracy of other parts of the model.

9.2 Accurate Costing in a Non-utilisation Model

Firstly, we show how we can produce accurate costing in a non-utilisation model. In such a model, we assume that the number of hours that a piece of equipment works in any time period (given its equipment type, age, and the operating hours of the mine) is pre-determined. This means that the working age of the equipment is determined by the length of time since it was bought, and this makes it quite simple to figure out when a piece of equipment crosses into a new age bracket.

Our basic usage variable is $\mathbf{x}_j^{k,l}$, which we define to be 1 if equipment j is owned in time period k, having been bought l periods before that, and 0 otherwise. We assume that age brackets are of constant size, denoted by B_0, and the operating hours of the mine are H^k for period k. In our model, we had $B_0 = 5000\,\text{h}$ and $H^k = 3500\,\text{h}$ (for all k). We use a_j^l to denote the availability of equipment j when it is l periods old. The availability is the proportion of the operating hours that the equipment works: since this is a non-utilistion model, all equipment is worked to the fullest in each period.

Assuming that $\mathbf{x}_j^{k,l} = 1$, the age bracket that equipment j lies in at the start of period k is

$$b_j^k(l) = \left\lfloor \frac{\sum_{h=k-l}^{k-1} a_j^{l+h-k} H^h}{B_0} \right\rfloor.$$

We now define $B_j^{k,l}$ to be the number of hours in period k that the equipment lies in age bracket $b_j^k(l) + 1$ (Fig. 9.3).

By assumption, equipment cannot jump age brackets twice in one period, so $a_j^l H^k - B_j^{k,l}$ is the number of hours spent in bracket $b_j^k(l)$. The equipment stays in bracket $b_j^k(l)$ throughout period k if and only if

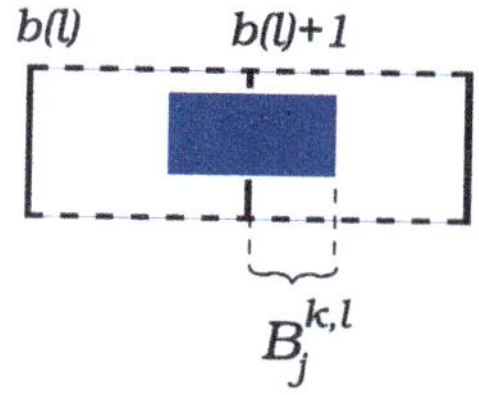

Fig. 9.3 Parameter B represents the amount of time the equipment spends in the next age bracket within the period. We write $b(l)$ as a simplified form of $b_j^k(l)$

$$\sum_{h=k-l}^{k} a_j^{l+h-k} H^h < (b_j^k(l) + 1) B_0,$$

and in this case, $B_j^{k,l} = 0$. Otherwise,

$$B_j^{k,l} = \sum_{h=k-l}^{k} a_j^{l+h-k} H^h - (b_j^k(l) + 1) B_0.$$

To calculate our cost, we let V_j^l be the variable cost (i.e. cost per operated hour) for equipment j that is in age bracket $b_j^k(l)$, and let D^k be the net present value factor associated with time period k. Then the cost of operating equipment j in time period k is

$$\sum_l D^k V_j^{b_j^k(l)} (a_j^l H^k - B_j^{k,l}) \mathbf{x}_j^{k,l} + \sum_l D^k V_j^{b_j^k(l)+1} B_j^{k,l} \mathbf{x}_j^{k,l}. \tag{9.1}$$

By adding this term to the objective function for every j and k, we produce an accurate assessment of the operating costs of the mine.

9.3 Utilisation and Cost Brackets in a Linear Model

In order to accurately assess the cost of operating a piece of equipment, we need to know the age of the equipment. In a utilisation model, the length of time since the equipment has been bought does not necessarily reflect the age of the equipment in terms of usage (although it does provide an upper bound).

We have a different usage variable for this model: we set the variable $\mathbf{x}_j^{k,l}$ to be 1 if equipment j is owned in time period k and starts this period in age bracket l, and 0 otherwise. This is different to the non-utilisation model because l now counts age brackets, rather than time periods. We also include a continuous utilisation variable, $\mathbf{f}_j^{k,l}$, which is the proportion of the operating hours that equipment j works in time period k, having started it in age bracket l.

We note that the age bracket that equipment j lies in at the start of period k is given by

$$\left\lfloor \sum_{h<k,l} \frac{H^h}{B_0} \mathbf{f}_j^{h,l} \right\rfloor.$$

Therefore, to ensure that only the correct $\mathbf{x}$ is nonzero, we include the constraints

$$l \leq \sum_{h<k,l} \frac{H^h}{B_0} \mathbf{f}_j^{h,l} \leq l + 1 - \varepsilon \tag{9.2}$$

when $\mathbf{x}_j^{k,l} = 1$, where ε is an arbitrarily small positive number. To apply these constraints only when $\mathbf{x}_j^{k,l} = 1$, we add or subtract a large multiple of $\mathbf{x}_j^{k,l}$ to the appropriate side, so that the constraints vanish when $\mathbf{x}$ is 0. This is shown in the following lemma.

Lemma 9.1 *To ensure that only the* $\mathbf{x}$ *with the correct age bracket is nonzero, we include the constraints*

$$\sum_{h<k,l} \frac{H^h}{B_0}\mathbf{f}_j^{h,l} \geq M(\mathbf{x}_j^{k,l} - 1) + l \qquad \forall \quad k, l, j \tag{9.3}$$

and

$$\sum_{h<k,l} \frac{H^h}{B_0}\mathbf{f}_j^{h,l} \leq l + 1 - \varepsilon + M(1 - \mathbf{x}_j^{k,l}) \qquad \forall \quad k, l, j \tag{9.4}$$

where M *is an arbitrarily large integer.*

We note that, in practice, M does not need to be arbitrarily large, but merely at least as big as the maximum number of age brackets.

The above results do not apply to pre-existing equipment, because such equipment already has a nonzero age at time 0. We can use similar constraints, but with a small modification to account for this. If we define $P(j)$ to be the age in hours of equipment j when the mine schedule starts, the age bracket that it lies in at the beginning of period k is now

$$\left\lfloor \frac{P(j) + \sum_{h<k,l} H^h \mathbf{f}_j^{h,l}}{B_0} \right\rfloor.$$

Constraints (9.3) and (9.4) can then be repeated by replacing the left-hand side with this expression.

9.4 Accurate Costing in a Utilisation Model

In a utilisation model, we do not assume that all equipment is worked to its fullest capacity. Instead, we allow ourselves to choose which pieces of equipment work, and for how long. This means that it is not so easy to calculate the true age of a piece of equipment.

As for the non-utilisation model, we need to know whether a given piece of equipment crosses into a new age bracket in the middle of a time period, and if so, for how long. To find this, it is not enough to know the age bracket of a piece of equipment—we must know its age in hours as well. With this in mind, we let $\boldsymbol{\beta}_j^{k,l}$ be a variable which is the amount of time, in hours, that equipment j operates in age bracket $l + 1$ in period k, given that it started that period in age bracket l (Fig. 9.4).

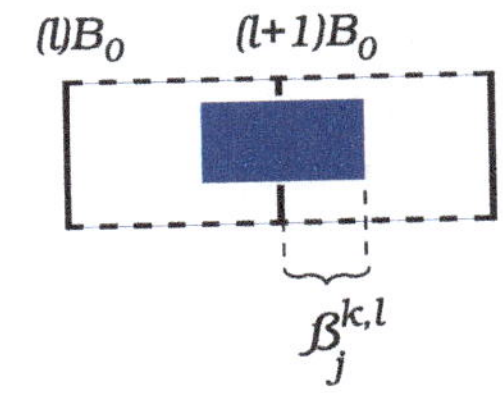

Fig. 9.4 Variable $\boldsymbol{\beta}$ represents the amount of time the equipment spends in the next age bracket within the period

We set a salvage variable, $\mathbf{s}_j^{k,l}$, to be 1 if equipment j is salvaged in time period k and starts that period in age bracket l, and 0 otherwise.

The first constraint on $\boldsymbol{\beta}$ is that it is non-negative:

$$\boldsymbol{\beta}_j^{k,l} \geq 0, \qquad \forall\, j, k, l. \tag{9.5}$$

Next, we restrict $\boldsymbol{\beta}$ to be 0 if the corresponding $\mathbf{x}$ is zero, indicating that the equipment was either not owned or in a different age bracket in the period:

$$\boldsymbol{\beta}_j^{k,l} \leq M\mathbf{x}_j^{k,l}, \qquad \forall\, j, k, l. \tag{9.6}$$

Now, the age in hours of equipment j at the end of time period k is given by

$$\sum_{h \leq k,l} H^h \mathbf{f}_j^{h,l}.$$

Since the start of the $(l+1)$th age bracket is $(l+1)B_0$ hours, we need $\boldsymbol{\beta}_j^{k,l}$ to be the difference between the age and $(l+1)B_0$, if this difference is positive. This gives a lower bound on $\boldsymbol{\beta}$, with the caveat that the corresponding $\mathbf{x}$ must be nonzero:

$$\boldsymbol{\beta}_j^{k,l} \geq \sum_{h \leq k,l} H^h \mathbf{f}_j^{h,l} - (l+1)B_0 - M(1 - \mathbf{x}_j^{k,l}) \qquad \forall\, j, k, l. \tag{9.7}$$

Together with the restriction that $\boldsymbol{\beta}$ is non-negative, this provides a tight lower bound for $\boldsymbol{\beta}$. However, the upper bound is slightly trickier. We cannot simply set $\boldsymbol{\beta}$ to be less than or equal to this difference, because it may be negative. We get around this by noting that we can tell when the equipment crosses over into the next age bracket, by observing the value of $\mathbf{x}_j^{k+1,l} + \mathbf{s}_j^{k+1,l}$. This is 1 if and only if the equipment starts period $k+1$ in age bracket l, and therefore has not shifted age brackets. This gives us the final constraint on $\boldsymbol{\beta}$:

$$\boldsymbol{\beta}_j^{k,l} \leq \sum_{h \leq k,l} H^h \mathbf{f}_j^{h,l} - (l+1)B_0 + M(1 - \mathbf{x}_j^{k,l} + \mathbf{x}_j^{k+1,l} + \mathbf{s}_j^{k+1,l}) \qquad \forall\, j, k, l. \tag{9.8}$$

For pre-existing equipment, the age in hours of the equipment is increased by $P(j)$. This propogates through to constraints (9.7) and (9.8) in the obvious way.

Now that we have set the value of $\boldsymbol{\beta}$, we can include it in the objective function. We define V_j^l and D^k as for the non-utilisation model. Accounting for the possibility of an age bracket shift gives us the operating cost

$$\sum_{j,k,l} D^k V_j^l H^k \mathbf{f}_j^{k,l} + \sum_{j,k,l} D^k (V_j^{l+1} - V_j^l) \boldsymbol{\beta}_j^{k,l}.$$

If we assumed that the cost is determined by the age bracket of the equipment at the start of the period, then the operating cost would have been just the first of these terms.

The addition in complexity to the model is not great, as there are the same number of $\boldsymbol{\beta}$ variables as $\mathbf{x}$ or $\mathbf{f}$ variables.

9.5 Accurate Utilisation in a Utilisation Model

Calculating the number of hours spent in the next age bracket also allows us to bound the utilisation variable $\mathbf{f}$ in a better way. Previous models merely assumed that the utilisation was bounded by the availability of the equipment, as determined at the start of the time period. In other words, we redefined a_j^l to be the availability of equipment j when it is in age bracket l, and applied the constraints $\mathbf{f}_j^{k,l} \le a_j^l$.

However, this suffers from the same problem that we encountered with costing: the equipment may cross into a new age bracket with significantly different availability during the time period. To solve this problem, we note that equipment j operates for $\boldsymbol{\beta}_j^{k,l}$ hours in age bracket $l+1$ during period k. The total number of hours spent in age bracket $l+1$, including non-operating time, must therefore be at least $\frac{1}{a_j^{l+1}} \boldsymbol{\beta}_j^{k,l}$. This gives us an upper bound for the total number of hours spent in age bracket l, which in turn gives an upper bound on the operating hours spent in age bracket l:

$$\mathbf{f}_j^{k,l} H^k - \boldsymbol{\beta}_j^{k,l} \le a_j^l \left(H^k - \frac{1}{a_j^{l+1}} \boldsymbol{\beta}_j^{k,l} \right).$$

Rearranging gives the modified bound on $\mathbf{f}$:

$$\mathbf{f}_j^{k,l} + \frac{1}{H^k} \left(\frac{a_j^l}{a_j^{l+1}} - 1 \right) \boldsymbol{\beta}_j^{k,l} \le a_j^l \qquad \forall\, j, k, l. \tag{9.9}$$

If $\boldsymbol{\beta}_j^{k,l} = 0$, this reduces to the original constraint.

Notation Index

A summary of the notation used in this chapter is as follows.

- $\mathbf{x}_j^{k,l}$ indicator variable if equipment j is onwed in time period k, having been purchased l periods before that.
- B_0 the constant size of age brackets (hours).
- H^k the operating hours of the mine (hours).
- $b_j^k(l)$ the age bracket that equipment j lies in at the start of period k.
- $B_j^{k,l}$ is the number of hours in period k that the equipment lies in age bracket $b_j^k(l)$.
- a_j^l the availability of equipment j aged l.
- V_j^l the variable cost of equipment j that is in age bracket $b_j^k(l)$.
- D^k the net present value discount factor associated with time period k.
- $\mathbf{f}_j^{k,l}$ the proportion of operating hours that equipment j works in time period k, having started l periods ago.
- ε an arbitrarily small continuous number.
- $P(j)$ the age in hours of equipment j when the mine schedule starts
- $\beta_j^{k,l}$ the amount of time, in hours, that equipment j operates in age bracket $l+1$ in period k.
- M an arbitrarily large integer.
- F_j the fixed cost (capital expense) of obtaining equipment type j.
- V_j^l the variable expense for equipment type e aged l in period k.

Acknowledgements Components of this chapter were presented in [1].

References

1. C. Burt, Y.B. Chan, Accurate costing in mixed integer utilisation mining models, in *18th IMACS World Congress - MODSIM09 International Congress on Modelling and Simulation*, ed. by R.S. Anderssen, R.D. Braddock, L.T.H. Newham (2009), pp. 74–80
2. A. Haidar, S. Naoum, R. Howes, J. Tah, Genetic algorithms application and testing for equipment selection. ASCE J. Constr. Eng. Manag. **25**(1), 32–38 (1999)
3. M. Hassan, G. Hogg, D. Smith, A construction algorithm for the selection and assignment of materials handling equipment. Int. J. Prod. Res. **23**(2), 381–392 (1985)

Chapter 10
Future Research Directions

Christina N. Burt and Louis Caccetta

This book addressed the equipment selection problem from an optimisation perspective. In the opening chapters of this book, we have provided context, background and definitions to the Equipment Selection and Allocation problem. In Chap. 1, we explored the elements of the problem that are important to consider for meaningful solutions. In Chap. 2, we extended this discussion to cover a basic introduction to methods for addressing the problem.

In Chap. 3, we summarised the literature on equipment selection and identified methodology that could be useful for future research directions. In particular, faster solutions could be achieved through the development of heuristics and approximation algorithms rather than taking an exact approach.

Problem 1 Derive an exact model for the complete problem, and focus on developing, implementing and testing efficient, near-optimal heuristic approaches to obtaining solutions.

Within that chapter we also established that decomposition methods had also not been well explored. Two potentially interesting approaches revolve around decomposition of the time periods.

The original version of this chapter was revised: Originator type has been changed and the chapter author names have been included. The erratum to this chapter is available at https://doi.org/10.1007/978-3-319-76255-5_11

C. N. Burt (✉)
Department of Mathematics and Statistics, The University of Melbourne,
Parkville, VIC, Australia
e-mail: c.burt@ms.unimelb.edu.au

L. Caccetta
Department of Mathematics and Statistics, Curtin University, Bentley, WA, Australia
e-mail: L.Caccetta@curtin.edu.au

C. N. Burt and L. Caccetta (eds.), *Equipment Selection for Mining: With Case Studies*, Studies in Systems, Decision and Control 150,
https://doi.org/10.1007/978-3-319-76255-5_10

Problem 2 Derive a decomposition approach that shares information of different time fidelity between short and long term planning. For example, solve the long term problem and inform it using partial solutions from a short term problem.

Problem 3 Derive a decomposition approach that exploits a rolling horizon utilising a more refined version of the problem in the current horizon, informed by crude versions in the long term.

Pre-existing equipment is common in mines when further equipment selection is required and it seems a short logical step to include this equipment in the process. A major contribution of our work begins with the consideration of such equipment.

In Chap. 4, we extended the important industry ratio match factor to be more flexible for mixed fleets (including pre-existing equipment) and multiple mining locations. We followed this work with with a detailed chapter on two case studies (Chap. 5), which we then utilised in three chapters presenting models of various complexity: multi-period equipment selection for single locations (Chap. 6), multi-period equipment selection for multiple locations (Chap. 7), and multi-period equipment selection accounting for utilisation of equipment (Chap. 8). Further extensions to this work are possible and warranted, particularly if the methodology recommendations from above are applied to extend the number of periods that can be efficiently considered, and also the number of mining locations.

A key assumption in these chapters is that the mine plan is deterministic. However, there are many aspects of the problem, such as elements of the cost function and the way the equipment ages as it is used, that is not deterministic. This highlights an important extension for future research.

Problem 4 Develop, implement and test an equipment selection model that is robust to uncertainty in the market and other stochastic elements, such as machine breakdown, maintenance and truck cycle times.

During the course of this work we identified several problems of interest that extend the work completed here. An important problem arising from this collection of work is the combined mining method selection and equipment selection problem.

Problem 5 Develop, implement and test an optimisation model for the combined mining method selection and equipment selection problem.

Given the influence of the pit optimisation on both mining method selection and equipment selection, one could consider mining method selection, pit optimisation and equipment selection performed together.

Problem 6 Develop, implement and test an optimisation model for the combined pit optimisation, mining method selection and equipment selection problem.

The effect of truck-loader interactivity on efficiency is not well understood in the industry, and has the potential to underpin the optimisation process. We discussed that queuing and bunching is important for the accurate estimation of cycle times. Sound queuing and bunching models would therefore be a valuable asset for the mining industry.

Problem 7 Develop a general queuing and bunching model for the estimation of truck cycle time.

Often the queuing and bunching of trucks is dependent on the types of equipment and also the fleet size. It may be possible that such a model can be integrated with an equipment selection model.

Problem 8 Develop an integrated queuing, bunching and equipment selection model.

The parameters assocated with queuing and bunching are only a small portion of the parameters that influence the decision making process and quality of the ESP solution.

Problem 9 Develop accurate computational models for estimation of critical parameters.

A potential drawback of heterogeneous equipment selection is the selection of multiple truck types in the optimal solutions. It is not known if such solutions are more costly if the mining schedule changes significantly. To study this we should compare retrospective case studies with these optimal solutions over a fully implemented mining schedule. Unfortunately data of this quality was not available for the research presented in this book.

Problem 10 Determine the relative costs associated with selecting multiple truck types over the full dynamic schedule of a surface mine.

The importance of accurate cost data cannot be over-emphasised. Equipment with different levels of accuracy in the data are incomparable. Equipment with poor data should not be included, as the results could be devastating if this equipment were selected and the costs increased significantly. This model may also be extended to multi-locations, although if the same concepts as were developed here are used this would require an exorbitant number of variables.

Problem 11 Extend the utilised cost based equipment selection model to multiple locations.

The future of surface mining is to have fully autonomous equipment. This is currently already happening with driverless trucks. These trucking operations have just the one truck type. In part this is for operational purposes, but as equipment gets more sophisticated lower maintenance costs favour homogeneous fleets—spare parts inventories can be maintained at lower costs. In this context, the ESP is really embedded in a larger Engineering Asset Management Problem.

Problem 12 Develop, implement and test a model for the Engineering Asset Management Problem that solves the homogeneous Equipment Selection Problem as a subproblem.

Erratum to: Equipment Selection for Mining: With Case Studies

Christina N. Burt and Louis Caccetta

Erratum to:
C. N. Burt and L. Caccetta, *Equipment Selection for Mining: With Case Studies*, Studies in Systems, Decision and Control 150, https://doi.org/10.1007/978-3-319-76255-5

In the original version of the book, the originator type "Authored Work" has to be now changed to "Edited Work" and the chapter author names have to be displayed in their respective chapters. The erratum book has been updated with the changes.

The updated online version of this book can be found at https://doi.org/10.1007/978-3-319-76255-5

C. N. Burt and L. Caccetta (eds.), *Equipment Selection for Mining: With Case Studies*, Studies in Systems, Decision and Control 150,
https://doi.org/10.1007/978-3-319-76255-5_11

The manufacturer's authorised representative in the EU is Springer Nature Customer Service Centre GmbH, Europaplatz 3, 69115 Heidelberg, Germany. If you have any concerns regarding our products, please contact ProductSafety@springernature.com

Printed and bound by CPI Group (UK) Ltd, Croydon, CR0 4YY
02/10/2026
02226885-0006